PERGAMON INTERNATIONAL LIBRARY
of Science, Technology, Engineering and Social Studies
*The 1000-volume original paperback library in aid of education,
industrial training and the enjoyment of leisure*
Publisher: Robert Maxwell, M.C.

Self-Consistent Fields in Atoms

SELECTED READINGS IN PHYSICS
GENERAL EDITOR: D. TER HAAR

SOME OTHER BOOKS IN RELATED SERIES

LINDSAY, R. B.
Men of Physics: Julius Robert Moyer—Prophet of Energy

LINDSAY, R. B.
Men of Physics: Lord Rayleigh—The Man and His Work

MEETHAM, A. R.
Basic Physics

SEEGER, R. J.
Men of Physics: Benjamin Franklin—New World Physics

SEEGER, R. J.
Men of Physics: Galileo Galilei His Life and Works

SEEGER, R. J.
Men of Physics: Josiah Willard

STEELS, H.
An O-Level Course in Physics—Volumes 2 and 3

TER HAAR, D.
Men of Physics: L. D. Landau, Volume 2

TER HAAR, D.
Problems in Undergraduate Physics

BILLINGTON, M. S.
Building Physics: Heat

BORN, M. and WOLF, E.
Principles of Optics. 5th Edition

SANDERS, J. H.
The Velocity of Light

McCARTHY, I. E.
Nuclear Reactions

The terms of our inspection copy service apply to all the above books. A complete catalogue of all books in the Pergamon International Library is available on request.

The Publisher will be pleased to receive suggestions for revised editions and new titles.

Self-Consistent Fields in Atoms

Hartree and Thomas–Fermi atoms

by

N. H. MARCH

*Department of Physics, Imperial College
of Science and Technology, London SW7*

PERGAMON PRESS

*Oxford · New York · Toronto
Sydney · Paris · Braunschweig*

QC173
M283

U. K.	Pergamon Press Ltd., Headington Hill Hall, Oxford OX3 0BW, England
U. S. A.	Pergamon Press Inc., Maxwell House, Fairview Park, Elmsford, New York 10523, U.S.A.
C A N A D A	Pergamon of Canada, Ltd., 207 Queen's Quay West, Toronto 1, Canada
A U S T R A L I A	Pergamon Press (Aust.) Pty. Ltd., 19a Boundary Street, Rushcutters Bay, N.S.W. 2011, Australia
F R A N C E	Pergamon Press SARL, 24 rue des Ecoles, 75240 Paris, Cedex 05, France
W E S T G E R M A N Y	Pergamon Press GmbH, 3300 Braunschweig, Postfach 2923, Burgplatz 1, West Germany

Copyright © 1975 **N. H. March**

First Edition 1975

Library of Congress Cataloging in Publication Data

March, Norman Henry, 1927–
Self-consistent fields in atoms.
(Selected readings in physics)
1. Atoms-Mathematical models. 2. Hartree–Fock approximation. I. Title.
QC173.M3633   539.7   74–8317
ISBN 0–08–017819–7
ISBN 0–08–017820–0 (flexicover)

Printed in Great Britain by A. Wheaton & Co., Exeter

Contents

PART II

Outline

This small Volume is concerned with atomic properties—energy levels, binding energies, how atoms scatter X-rays, what magnetic properties they have and so on.

At the heart of this description of atoms is the Hartree field concept. This is a one-electron description and, in consequence, it does not deal at all with certain collective properties arising from electron–electron correlations which exist due to the Coulomb repulsions which occur between electrons. Therefore in Chapter 8 we transcend the self-consistent field concept and discuss in fairly general terms the effects of electron correlation and the way in which certain atomic properties require us to incorporate collective effects into the theory. The Volume concludes with a discussion of relativistic corrections which must be included in the theory of heavy atoms.

It is a pleasure to thank Drs. J. M. Titman and G. E. Kilby for useful comments on some parts of the MS.

PART I

CHAPTER 1

Central field wave functions and angular momentum operators

1.1. Introduction

One of the earliest successes of the quantum theory was of course Bohr's calculation of the energy level spectrum of the hydrogen atom. Though derived from a picture that was too classical in that electrons were still thought of as orbiting round the proton in sharply defined classical paths, this level spectrum was found later to agree with that calculated from wave mechanics. In particular, if we solve the Schrödinger wave equation

$$\nabla^2\psi + \frac{8\pi^2 m}{h^2}[E - V(r)]\psi = 0, \qquad (1.1)$$

for the wave functions Ψ and energies E with potential energy $V(r) = -e^2/r$ appropriate to the Coulomb attraction between a proton and an electron at distance r from it, then the Bohr level spectrum[†]

$$E_n = -\frac{e^2}{2n^2 a_0}, \qquad n = 1, 2 \ldots, \qquad (1.2)$$

is regained. Here a_0 is the first Bohr radius for hydrogen, given by $a_0 = h^2/4\pi^2 me^2 = 0.529$ Å. The energy e^2/a_0 is in magnitude equal to 27 eV and thus the ground state energy of the hydrogen atom ($n = 1$) is $-13 \cdot 5$ eV. In the above treatment the proton is assumed fixed;

[†] The reader interested in the derivation of eqn. (1.2) should consult Appendix 2.1.

3

a good approximation since its mass is about 2000 times as great as the electron mass m. In fact, the motion of the proton can be incorporated as a correction when necessary: the problem is again exactly soluble (see, for example, Pauling and Wilson, 1935).

Unfortunately, when the above treatment is generalized to apply to the next simplest atom He, the problem is already so complex that no exact solution has yet been found. For atoms other than hydrogen, we are, therefore, led to simplify the problem by giving each electron its own personal wave function, which we will require to satisfy a Schrödinger equation like eqn. (1.1), but in which the pure Coulomb field appropriate to hydrogen must now be modified.

1.2. Angular momentum and central fields

Accepting that each electron in an atom can usefully be described by the Schrödinger eqn. (1.1), we need, of course, to know the potential energy $V(\mathbf{r})$ representing the force field in which this electron moves.

Fortunately, without exploring the detailed form of $V(\mathbf{r})$, we can make immediate progress with the single assumption that $V(\mathbf{r})$ is spherically symmetrical, that is $V(\mathbf{r}) \equiv V(|\mathbf{r}|) = V(r)$. This so-called central field assumption leads, in classical mechanics, immediately to one of Kepler's laws of planetary motion, namely that equal areas are swept out by the line joining the planet under discussion to the Sun in equal times. This follows from the constancy of the orbital angular momentum, defined such that the vector $\mathbf{L}$ is given in terms of position vector $\mathbf{r}$ and momentum vector $\mathbf{p}$ by

$$\mathbf{L} = \mathbf{r} \times \mathbf{p}. \tag{1.3}$$

If we now form the rate of change of $\mathbf{L}$ with time, $\dot{\mathbf{L}}$, then it follows immediately from equation (1.3) that

$$\dot{\mathbf{L}} = \dot{\mathbf{r}} \times \mathbf{p} + \mathbf{r} \times \dot{\mathbf{p}}. \tag{1.4}$$

For a particle of mass m, we evidently have $\mathbf{p} = m\dot{\mathbf{r}}$ and thus the first term on the right-hand side of eqn. (1.4) vanishes in general since it involves $\dot{\mathbf{r}} \times \dot{\mathbf{r}}$.

To obtain the second term on the right-hand side, we use the classical Newtonian equation $\mathbf{F} = \dot{\mathbf{p}}$, where the force $\mathbf{F}$ is related to the potential energy $V(\mathbf{r})$ by

$$\mathbf{F} = -\operatorname{grad} V. \tag{1.5}$$

Since $V = V(r)$ in central field problems it follows from eqn. (1.5) that the force is purely radial, i.e. along the direction of the unit vector $(\mathbf{r}/r)$. Again, therefore, since $\mathbf{r} \times \mathbf{r}$ now appears in eqn. (1.4), the second term on the right-hand side is also zero and hence $\dot{\mathbf{L}} = 0$. Thus, for classical motion in a central field, the angular momentum vector $\mathbf{L}$ is a constant, independent of time.

When we solve the Schrödinger eqn. (1.1) for a central field, we shall find that the angular momentum again plays a central role in the theory. But now, of course, the components of $\mathbf{L}$ given by eqn. (1.3) are operators since

$$\mathbf{p} \to \frac{\hbar}{i} \nabla, \quad \hbar = \frac{h}{2\pi}, \tag{1.6}$$

in Schrödinger wave mechanics. We shall return to the properties of the angular momentum operators when we have solved the Schrödinger equation for the central field wave functions.

1.3. Shapes of atomic wave functions

The natural coordinate system in which to work when $V(\mathbf{r}) \equiv V(r)$ is that of spherical polar coordinates r, θ, ϕ, related to Cartesian coordinates (see Fig. 1.1) by

$$\left. \begin{aligned} x &= r \sin \theta \cos \phi \\ y &= r \sin \theta \sin \phi \\ z &= r \cos \theta \end{aligned} \right\} \tag{1.7}$$

with $0 \leqslant r < \infty$, $0 \leqslant \theta \leqslant \pi$, $0 \leqslant \phi \leqslant 2\pi$. The Laplacian operator ∇^2 in the Schrödinger eqn. (1.1) then takes the form (see, for example,

Rutherford, 1940)

$$\nabla^2 = \frac{1}{r^2}\,\frac{\partial}{\partial r}\left(r^2\,\frac{\partial}{\partial r}\right) + \frac{1}{r^2\sin\theta}\,\frac{\partial}{\partial\theta}\left(\sin\theta\,\frac{\partial}{\partial\theta}\right)$$
$$+ \frac{1}{r^2\sin^2\theta}\,\frac{\partial^2}{\partial\phi^2}\,. \tag{1.8}$$

Even though V depends solely on r, and not on θ and ϕ, this is not true of the wave functions $\psi(\mathbf{r})$ in general; the wave functions can have

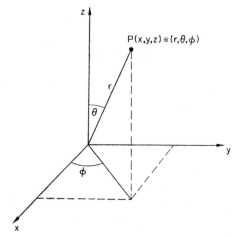

FIG. 1.1. Spherical polar coordinate system. Explicit relations between (x, y, z) and (r, θ, ϕ) are given in eqn. (1.7)

a lower symmetry than the potential energy V in the Schrödinger equation.

However, we can still exploit the fact that V is spherically symmetric, and thus is independent of θ and ϕ, as we can see by seeking a solution for ψ of the form

$$\psi = R(r)\,S(\theta, \phi). \tag{1.9}$$

It is evident that $R(r)$, describing the radial extent of the electronic wave function, must depend on the details of the potential energy $V(r)$, whereas we shall see that the shape of the atomic wave functions, described by $S(\theta, \phi)$, can be obtained once and for all in central field problems. Using eqns. (1.1), (1.8) and (1.9), and dividing both sides of the resulting Schrödinger equation by (RS/r^2) it follows that

$$\frac{1}{R} \frac{\partial}{\partial r} \left(r^2 \frac{\partial R}{\partial r} \right) + \frac{1}{S \sin \theta} \frac{\partial}{\partial \theta} \left(\sin \theta \frac{\partial S}{\partial \theta} \right)$$

$$+ \frac{1}{S \sin^2 \theta} \frac{\partial^2 S}{\partial \phi^2}$$

$$+ \frac{2m}{\hbar^2} [E - V(r)] r^2 = 0. \qquad (1.10)$$

The first and fourth terms on the left-hand side of eqn. (1.10) are evidently functions solely of the radial distance r, while the other two terms are functions of the angular variables θ and ϕ alone. Therefore, if eqn. (1.10) is to hold for *all* r, θ and ϕ, we must have separately that

$$\frac{1}{R} \frac{\partial}{\partial r} \left(r^2 \frac{\partial R}{\partial r} \right) + \frac{2m}{\hbar^2} [E - V(r)] r^2 = \text{constant}$$

$$= \mu, \text{ say} \qquad (1.11)$$

and

$$\frac{1}{\sin \theta} \frac{\partial}{\partial \theta} \left(\sin \theta \frac{\partial S}{\partial \theta} \right) + \frac{1}{\sin^2 \theta} \frac{\partial^2 S}{\partial \phi^2} = -\mu S. \qquad (1.12)$$

It is clear now that while we can solve eqn. (1.12) generally, and hence discuss the *shapes* of atomic wave functions, we cannot deal with the radial wave function $R(r)$ without explicitly choosing $V(r)$. Nevertheless, before studying the solutions $S(\theta, \phi)$ of eqn. (1.12), it is of interest to examine equation (1.11) a little further. To do so, we rewrite it in a way which makes it clear that it is, in essence, a one-dimensional Schrödinger equation, but with the potential $V(r)$ modified in a definite manner. Thus, we find, after a little manipulation,

$$\frac{\partial^2}{\partial r^2} (rR) + \frac{2m}{\hbar^2} \left[E - V(r) - \frac{\hbar^2}{2m} \frac{\mu}{r^2} \right] (rR) = 0 \qquad (1.13)$$

which is indeed a Schrödinger equation for the product rR, with an 'effective' potential energy

$$V_{\text{eff}}(r) = V(r) + \frac{\hbar^2}{2m} \frac{\mu}{r^2}.$$ (1.14)

To understand the origin of the term modifying $V(r)$ in eqn. (1.14), let us return to the elementary classical mechanical case of an electron moving in a circular orbit of radius r round a fixed proton. If the tangential velocity of the electron is v, then the Newtonian equation of motion yields for the radial force F

$$F = \frac{mv^2}{r} = \frac{(mvr)^2}{mr^3} = \frac{L^2}{mr^3}$$ (1.15)

where we have used the fact that the orbital angular momentum L equals mvr in this case, as follows from the general definition (1.3). Since potential energy and force are related by eqn. (1.5), we see that corresponding to eqn. (1.15) there is a potential energy term of the form $L^2/2mr^2$. Thus the interpretation of the separation constant in eqns. (1.11) and (1.12) appears from this argument to be that

$$\mu = \frac{L^2}{\hbar^2}.$$ (1.16)

We shall see below from a fully quantum-mechanical argument that this is indeed correct, and furthermore, L^2 is quantized, with allowed values $l(l+1)\hbar^2$ where $l = 0, 1, 2$, etc. The basic unit in which angular momentum is measured in quantum mechanics is $\hbar$.

The effective potential energy $V_{\text{eff}}(r)$ is indicated schematically in Fig. 1.2 for the case when $V(r)$ is the Coulomb potential $-e^2/r$ and it can be seen that, except for $l = 0$, when $V_{\text{eff}} = V(r)$, the modification due to the 'centrifugal potential energy' is major at sufficiently small r. We must therefore except the behaviour of atomic wave functions near the nucleus to be crucially dependent on the angular momentum of the state in question.

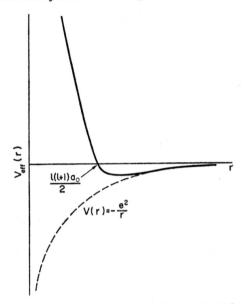

Fɪɢ. 1.2. Schematic form of $V_{\text{eff}}(r)$ as defined in eqn. (1.14) for $V(r) = -e^2/r$ and $l \neq 0$.

(a) *Dependence of wave functions on ϕ and magnetic quantum number m*

Though eqn. (1.12) will be recognized as it stands by mathematicians, let us take the further step of writing

$$S(\theta, \phi) = \Theta(\theta)\,\Phi(\phi). \tag{1.17}$$

Then, we immediately find that eqn. (1.12) separates, the new equations for Θ and Φ being, after dividing through by $[\Theta\Phi/\sin^2\theta]$,

$$\frac{1}{\Phi}\frac{\partial^2\Phi}{\partial\phi^2} = -m^2 \tag{1.18}$$

and

$$\frac{\sin\theta}{\Theta}\frac{\partial}{\partial\theta}\left(\sin\theta\frac{\partial\Theta}{\partial\theta}\right) + \mu\sin^2\theta = m^2. \tag{1.19}$$

The form of the separation constant adopted in eqn. (1.18), namely $-m^2$, where m is taken to be real, has anticipated the fact that we want periodic solutions for $\Phi(\phi)$, simply because, adding multiples of 2π to ϕ (see Fig. 1.1) could not change the physical properties of the system. Therefore the solution of eqn. (1.18) which we adopt is

$$\Phi(\phi) = \exp(im\phi) \tag{1.20}$$

and the requirement that $\Phi(\phi) = \Phi(\phi+2\pi n)$, where n is an integer, yields

$$m = 0, \quad \pm 1, \quad \pm 2, \quad \text{etc.} \tag{1.21}$$

m is the so-called magnetic quantum number, determining the way the energy levels of an atom split in an applied magnetic field. For our purposes, however, it is important to connect m with the properties of the angular momentum operator $\mathbf{L}$. Evidently, from eqn. (1.3) we can write the z component of angular momentum L_z as

$$L_z = xp_y - yp_x \tag{1.22}$$

and hence the quantum-mechanical operator has the form

$$L_z = \frac{\hbar}{i}\left[x\frac{\partial}{\partial y} - y\frac{\partial}{\partial x}\right] \tag{1.23}$$

which follows immediately from (1.6) and (1.22).

Transforming eqn. (1.23) to spherical polar coordinates, using the relations (1.7), we find

$$L_z = \frac{\hbar}{i}\frac{\partial}{\partial \phi}. \tag{1.24}$$

Hence, from eqn. (1.20) it follows that

$$L_z\Phi = m\hbar\Phi \tag{1.25}$$

and therefore $\Phi(\phi)$ is an *eigenfunction* of the operator L_z, with eigenvalue $m\hbar$. Thus, since m is integral, L_z is quantized in units of $\hbar$. This completes then the discussion of the ϕ dependence of the central-field wave functions and we turn next to the θ dependence.

(b) *θ-dependence of wave functions and azimuthal quantum number l*

To solve eqn. (1.19), it is convenient to make the substitutions

$$x = \cos \theta, \quad \Theta(\theta) = P(x). \tag{1.26}$$

Then we find that $P(x)$ satisfies the equation

$$\frac{d}{dx}\left[(1-x^2)\frac{dP}{dx}\right] + \left[\mu - \frac{m^2}{1-x^2}\right]P = 0 \tag{1.27}$$

and x goes from $-1 \leqslant x \leqslant +1$, corresponding to $0 \leqslant \theta \leqslant \pi$.

Investigation of eqn. (1.27) shows that, for a general value of μ, the solutions become infinite at $x = \pm 1$, as might be anticipated from the presence of the term $m^2/(1-x^2)$. These solutions are therefore not leading to well-behaved wave functions ψ and are not physically acceptable. Only in the special cases when $\mu = l(l+1)$, $l = 0$, 1, 2, etc, $|m| \leqslant l$, does the equation for P possess one solution which is everywhere finite. These solutions are, in fact the associated Legendre functions of order l (see, for instance, Sneddon, 1951b). They can be defined in the following way:

$$P_l^m(x) = (1-x^2)^{|m|/2} \frac{d^{|m|}}{dx^{|m|}} P_l(x), \quad |m| < l, \quad l = 0, 1, 2, \text{ etc.} \tag{1.28}$$

where the Legendre polynomials $P_l(x)$ are given by

$$P_l(x) = \frac{1}{2^l l!} \frac{d^l}{dx^l} [(x^2-1)^l], \quad l = 0, 1 \dots \tag{1.29}$$

The Legendre polynomials therefore satisfy eqn. (1.27) with $\mu = l(l+1)$ and $m = 0$, whereas the associated Legendre functions, satisfying eqn. (1.27) for general m, have the form $(1-x^2)^{|m|/2} \times$ [a real polynomial of degree $(l-|m|)$]. The first four Legendre polynomials are:

$$P_0(x) = 1 \qquad\qquad P_1(x) = x$$
$$P_2(x) = \tfrac{1}{2}(3x^2-1) \qquad P_3(x) = \tfrac{1}{2}(5x^3-3x) \tag{1.30}$$

and in general the even polynomials $P_{2s}(x)$ are even in x, while the odd polynomials $P_{2s+1}(x)$ are themselves odd in x. The first few associated Legendre functions are

$$
\left.
\begin{aligned}
P_1^{\pm 1}(x) &= (1-x^2)^{\frac{1}{2}} \\[4pt]
P_2^{\pm 1}(x) &= (1-x^2)^{\frac{1}{2}}\, 3x \\[4pt]
P_3^{\pm 1}(x) &= (1-x^2)^{\frac{1}{2}}\, \tfrac{3}{2}[5x^2-1]
\end{aligned}
\right\}
$$

and

$$
\left.
P_2^{\pm 2}(x) = 3(1-x^2).
\right\}
\tag{1.31}
$$

In conformity with a general property of solutions of the Schrödinger equation discussed in Appendix 1.1, the associated Legendre functions are orthogonal over the interval $-1 \leqslant x \leqslant 1$, that is

$$
\int_{-1}^{1} P_l^m(x)\, P_{l'}^m(x)\, dx = 0, \qquad l' \neq l. \tag{1.32}
$$

For $l = l'$ we have

$$
\int_{-1}^{+1} \{P_l^m(x)\}^2\, dx = \left[\frac{2}{2l+1} \frac{(l+|m|)!}{(l-|m|)!} \right] \tag{1.33}
$$

which allows us to define normalized associated Legendre functions, apart from an arbitrary phase factor.

(c) *Eigenfunctions of L^2 and L_z*

The complete solutions $S(\theta, \phi)$ for the cases $\mu = l(l+1)$, $l = 0, 1, 2$, etc, and $|m| \leqslant l$ are in fact the spherical harmonics $Y_{lm}(\theta, \phi)$. Hence we can write

$$
Y_{lm}(\theta, \phi) = N_{lm} P_l^m(\cos\theta)\, \Phi_m(\phi). \tag{1.34}
$$

There is no universally accepted definition of the normalization factor N_{lm} but if we require

$$
\int_0^{2\pi} d\phi \int_0^{\pi} d\theta\, \sin\theta\, Y_{lm}^*(\theta, \phi)\, Y_{l'm'}(\theta, \phi) = 0 \tag{1.35}
$$

unless $l = l'$ and $m = m'$ and is unity in this case, then N_{lm} is determined by

$$\frac{1}{|N_{lm}|^2} = \left[\frac{4\pi(l+|m|)!}{(2l+1)(l-|m|)!} \right].$$ (1.36)

As two examples, we have

$$\left. \begin{array}{l} Y_{10} = \left(\dfrac{3}{4\pi} \right)^{\frac{1}{2}} \cos\theta \\[4mm] Y_{11} = \left(\dfrac{3}{8\pi} \right)^{\frac{1}{2}} \sin\theta\, e^{i\phi} \end{array} \right\}.$$ (1.37)

We now return to the discussion of the angular momentum operator L^2. In spherical polar coordinates we have, corresponding to eqn. (1.24) for L_z the results

$$L_x = i\hbar \left[\sin\phi \frac{\partial}{\partial\theta} + \cot\theta\cos\phi \frac{\partial}{\partial\phi} \right]$$ (1.38)

and

$$L_y = i\hbar \left[-\cos\phi \frac{\partial}{\partial\theta} + \cot\theta\sin\phi \frac{\partial}{\partial\phi} \right].$$ (1.39)

Forming $L^2 = L_x^2 + L_y^2 + L_z^2$ we find from eqns. (1.24), (1.38) and (1.39)

$$L^2 = -\hbar^2 \left\{ \frac{1}{\sin\theta} \frac{\partial}{\partial\theta} \left(\sin\theta \frac{\partial}{\partial\theta} \right) + \frac{1}{\sin^2\theta} \frac{\partial^2}{\partial\phi^2} \right\}.$$ (1.40)

Returning to eqn. (1.12) satisfied by $S = Y_{lm}(\theta, \phi)$ when $\mu = l(l+1)$, we see that the operator on the left-hand side acting on S is in fact $-L^2/\hbar^2$ from eqn. (1.40). Thus, we find the result

$$L^2 S \equiv L^2 Y_{lm}(\theta, \phi) = l(l+1)\hbar^2 Y_{lm}(\theta, \phi).$$ (1.41)

But we know already that $\Phi(\phi) = e^{im\phi}$ is an eigenfunction of L_z with eigenvalue $m\hbar$ and hence we conclude that the spherical harmonic $Y_{lm}(\theta, \phi)$ is a simultaneous eigenfunction of L^2 and L_z with eigen-values $l(l+1)\hbar^2$ and $m\hbar$ respectively. The former result confirms the earlier deduction in eqn. (1.16), made on classical grounds.

Thus the central field wave functions correspond to sharp values of L^2 and L_z and this is the quantal analogue of the classical result that the angular momentum is constant for a central field of force. All three components of **L** can be precisely specified at once in classical theory.

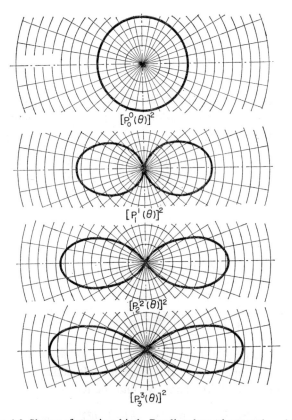

$[P_0^0(\theta)]^2$

$[P_1^1(\theta)]^2$

$[P_2^2(\theta)]^2$

$[P_3^3(\theta)]^2$

FIG. 1.3. Shapes of atomic orbitals. Reading down the page the orbitals are of s, p, d and f types ($l = 0, 1, 2$ and 3) respectively, the squares of the associated Legendre functions being plotted (cf. Pauling and Wilson, 1935).

However, in general only L^2 and L_z are specified in the quantum mechanical central field problem, since $Y_{ml}(\theta, \phi)$ is *not* an eigenfunction of L_x and L_y,[†] except when $l = 0$.

We have dealt with the problem of the shape of atomic wave functions at length here because these results are applicable to *every* central field problem. The shapes of these orbitals are of considerable importance, especially when one wishes to describe atoms bound in a molecule or a crystal (see, for example, Coulson 1961). We show in Fig. 1.3 a few examples of the shapes of particular wave functions, the spectroscopic notation *s*, *p*, *d*, *f*, *g*, *h* (historically, sharp, principal, diffuse, fundamental while *g*, *h* ... follow alphabetically after *f*) corresponding to $l = 0, 1, 2, 3$, etc, respectively. The *s* states are spherically symmetrical, while the *p*, *d* etc. states all show angularity. It is difficult to overemphasize the role of angular momentum (not only orbital, but also spin which we discuss later) in atomic theory (for an advanced discussion, see, for example, Condon and Shortley, 1951).

Problems

1. In classical theory use the constancy of the angular momentum in a central field (see section (1.2)) to prove Kepler's Law of planetary motion that equal areas are swept out in equal times.

2. Using classical mechanics, and treating a plane orbit using polar coordinates r and θ such that $x = r \cos \theta$, $y = r \sin \theta$, show that the transverse acceleration of the planet is given by $2\dot{r}\dot{\theta} + r\ddot{\theta}$, while the radial acceleration is $\ddot{r} - r\dot{\theta}^2$. Hence show explicitly that $r^2\dot{\theta}$ is a constant, independent of time, for a central field.

In a gravitational field of force F given by

$$F = -GmM/r^2$$

where G is the gravitational constant, m is the mass of the planet and M the mass of the Sun, show:

(i) That the differential equation of the planetary orbit is

$$\frac{d^2u}{d\theta^2} + u = \frac{GM}{h^2}, \quad h = r^2\dot{\theta}$$

with $u = 1/r$.

[†] The choice of the direction of the polar axis that distinguishes L_z from L_x and L_y is arbitrary, however.

(ii) That the conic section

$$\frac{1}{r} = C \cos \theta + \frac{\gamma}{h^2}, \quad \gamma = GmM,$$

satisfies this equation.

(iii) That for an elliptical orbit, the Kepler law

$$T^2 \propto a^3$$

holds, where T is the period of the orbital motion and a is the semimajor axis of the ellipse.

(iv) That the total energy E can be written as

$$E = -\frac{GmM}{2a}.$$

3. Given that, in cartesian coordinates, the three degenerate wave functions corresponding to the first excited state of a three-dimensional isotropic harmonic oscillator are (in suitable units)

$$xe^{-r^2}, \quad ye^{-r^2}, \quad ze^{-r^2},$$

form three central field wave functions. What are the eigenvalues of L^2 and L_z for the three cases?

4. You are given the function

$$\left[\frac{x^4 + y^4 + z^4}{r^4} - \frac{3}{5} \right].$$

Is this an eigenfunction of L^2? If so, find the eigenvalue. Does it correspond to a definite value of L_z?

5. Prove that the components of orbital angular momentum **L** in quantum mechanics obey the commutation relations

$$L_x L_y - L_y L_x \equiv [L_x, L_y] = i\hbar L_z$$

or in vector form

$$\mathbf{L} \times \mathbf{L} = i\hbar \mathbf{L}.$$

Show also that

$$[L_z, L^2] = 0.$$

CHAPTER 2

Concept of self-consistent field

2.1. Size of atomic wave functions

Having dealt rather fully with the shapes of central field wave functions in the previous chapter, we must treat next their spatial extent. We have then to solve the radial wave eqn. (1.11) and to do so it is clear that we must specify the potential energy $V(r)$. Actually, for realistic forms of $V(r)$ in multi-electron atoms, this equation must be solved numerically: an easy task on a modern electronic computer. However, for the hydrogen atom, with $V(r) = -e^2/r$, the radial wave functions can be obtained analytically. Because these radial wave functions apply only to hydrogen, or to one-electron hydrogen-like ions, they do not have the same generality as the angular functions we discussed previously. Therefore, we relegate the detailed solution of the Coulomb field case to Appendix 2.1.

The radial wave functions do have some common properties, nevertheless, independent of the detailed potential $V(r)$, and in particular, for $l = 0$, $m = 0$, the radial wave functions $R_{n, l=0}(r)$, classified by the principal quantum number n (see eqn. (1.2)) have $n-1$ radial nodes. The forms for $n = 1$, 2 and 3 for the s states $l = 0$ are shown in Fig. 2.1 (a), while two p state ($l = 1$) functions are plotted in Fig. 2.1 (b). It is shown in Appendix 2.1 that the ground-state radial wave function for hydrogen has the (unnormalized) form (see eqns. (A2.1.17) and (A2.1.18) with $n = 1$, $l = 0$)

$$R_{10}(r) = \exp\left(-\frac{r}{a_0}\right). \tag{2.1}$$

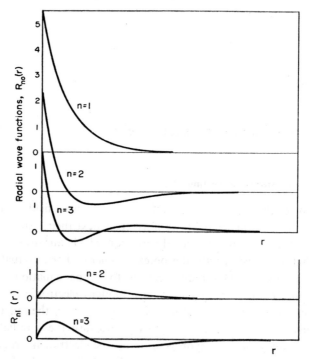

FIG. 2.1. (a) Form of radial wave functions $R_{n0}(r)$ for the hydrogen
atom, for $n = 1$, 2 and 3. (b) Form of radial wave functions for two p
states ($l = 1$), corresponding to principal quantum numbers $n = 2$
and 3.

The s wave function associated with the (degenerate)[†] first excited
state is, explicitly,

$$R_{20}(r) = \left[2 - \frac{r}{a_0} \right] \exp\left(-\frac{r}{2a_0} \right), \qquad (2.2)$$

its radial node evidently occurring at twice the Bohr radius. The position of this node will, of course, move, when we change the potential
energy from $-e^2/r$: that is the radial extent of the wave functions will

---

[†] We remind the reader that a state is said to be degenerate if there is more than
one wave function corresponding to the same energy.

depend on the potential energy $V(r)$. A further property which is general is that $R_{nl}(r) \sim r^l$ for sufficiently small r, the centrifugal potential energy term in eqn. (1.14) dominating the Coulomb term for $l \neq 0$ (see Fig. 1.2). This property can be seen from Fig. 2.1 for s and p states.

In multi-electron atoms, we need to decide how the field in which a given electron moves is to be set up before we can tackle the problem of the spatial extent of the electron cloud. It is natural enough to try to answer this question by looking at the next simplest atom, helium, with two electrons. As we remarked above, this problem has not yet been solved exactly.

2.2. Wave equation for helium atom

The Schrödinger equation for this case must evidently involve kinetic energy terms for two electrons, with positional coordinates $\mathbf{r}_1$ and $\mathbf{r}_2$ (again we assume the nucleus to be fixed). In quantum mechanics, the kinetic energy T, like any other dynamical variable, has an operator associated with it, and explicitly, since $T = p^2/2m$, we find from eqn. (1.6) that

$$T_{\mathrm{op}} = -\frac{\hbar^2}{2m} \nabla^2. \qquad (2.3)$$

Then the classical statement

$$H = E \qquad (2.4)$$

where H is the Hamiltonian[†] of the system becomes

$$H\Psi = E\Psi \qquad (2.5)$$

which is the Schrödinger equation, written explicitly for a single particle in eqn. (1.1).

For the two electrons in He, assuming again that the nucleus is so massive relative to the electrons that we can regard it as fixed, we have the Hamiltonian operator

---

† The Hamiltonian is, of course, expressed in terms of momenta and coordinates, i.e. $H = (p^2/2m) + V(r)$ for a particle of mass m in potential energy $V(r)$.

$$H_{\text{op}} = -\frac{\hbar^2}{2m}\triangle_1^2 - \frac{\hbar^2}{2m}\nabla_2^2 - \frac{Ze^2}{r_1} - \frac{Ze^2}{r_2} + \frac{e^2}{r_{12}} \qquad (2.6)$$

where reference can be made to Fig. 2.2 to see that $\mathbf{r}_1$ and $\mathbf{r}_2$ denote the position vectors joining electrons 1 and 2 to the nucleus, while r_{12} is the interelectronic distance. For helium itself, the charge Ze on the nucleus is, of course, simply $2e$, but the generalization to the case of 2 electrons in the field of a nuclear charge Ze effected in eqn. (2.6), is useful, as we shall see below.

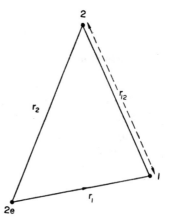

Fig. 2.2. Coordinate system for helium atom used in eqn. (2.6). r_{12} is the interelectronic distance,

The essential new feature in eqn. (2.6) is the presence of the Coulombic repulsion between the two electrons. If we could neglect e^2/r_{12}, then the Hamiltonian (2.6) would be simply a sum of two terms H_1 and H_2, the first being simply a function of the coordinates of electron 1, namely

$$-\frac{\hbar^2}{2m}\nabla_1^2 - \frac{Ze^2}{r_1},$$

with H_2 a similar term involving only $\mathbf{r}_2$. It is clear that the total energy is then a sum of two parts, and if each electron were in its ground state, we could use the Bohr formula (1.2), which, when generalized to a nucleus of charge Ze reads

$$E_n = -\frac{Z^2}{2n^2}\frac{e^2}{a_0}, \tag{2.7}$$

to see that the ground state energy of the He atom, in this approximation, is simply twice $E_1 = -2e^2/a_0$: that is -4 atomic units (au), an atomic unit of energy being $e^2/a_0 = 27$ eV. Comparing this with the experimental value of $-2\cdot9$ au for He, the effect of the Coulomb repulsion term e^2/r_{12} is seen to be very substantial and its neglect is not therefore permissible.

One further remark about the Hamiltonian. (2.6) must be made, when the interelectronic repulsion e^2/r_{12} is omitted. The wave function $\Psi(\mathbf{r}_1\mathbf{r}_2)$ obtained from

$$H_{\text{op}}\,\Psi(\mathbf{r}_1\mathbf{r}_2) = E\Psi(\mathbf{r}_1\mathbf{r}_2) \tag{2.8}$$

with H_{op} given by (2.6) with e^2/r_{12} omitted is clearly a product, as can be directly proved from (2.8). But it is simplest to argue that $\Psi^2(\mathbf{r}_1\mathbf{r}_2)$ is a probability density, and that if the electrons move independently, as they do when the term e^2/r_{12} is neglected, then probabilities of independent events are multiplicative and we can write

$$\Psi^2(\mathbf{r}_1\mathbf{r}_2) = \psi_n^2(\mathbf{r}_1)\,\psi_m^2(\mathbf{r}_2). \tag{2.9}$$

The wave function of the two-electron system is then a product of the wave functions $\psi_n(\mathbf{r}_1)$ and $\psi_m(\mathbf{r}_2)$ which are eigenfunctions of H_1 and H_2 respectively.

This product form

$$\Psi(\mathbf{r}_1\mathbf{r}_2) = \psi_n(\mathbf{r}_1)\,\psi_m(\mathbf{r}_2)^\dagger \tag{2.10}$$

is very basic for the ensuing discussion. We want to emphasize that, whereas eqn. (2.10) has been derived on the basis of the neglect of the e^2/r_{12} term in eqn. (2.6), leading inevitably to ψ_n and ψ_m in eqn. (2.10) as hydrogenic-like wave functions, with the charge on the nucleus simply $2e$, we can argue physically that a better approximation to the wave function satisfying (2.8) with the full Hamiltonian (2.6) including e^2/r_{12} would be obtained by reducing the nuclear charge somewhat.

---

† This function is only symmetric in the interchange of coordinates $\mathbf{r}_1$, and $\mathbf{r}_2$ when $n = m$. For the ground state of He, this is in fact the case. A fuller discussion of symmetry is deferred to Chapter 7.

The argument would go as follows. One of the two electrons, say 1, will obviously spend part of its time nearer to the nucleus than electron 2. When we try to write down a wave function for electron 2, we ought not, therefore, to suppose it to be acted on by the field of a nucleus of charge $2e$, but to be moving in a somewhat weaker field, since the electron 1 is screening it from the full attractive power of the nucleus. The electron 1, in other words, ought to be taken into account when we are calculating the wave function of the second electron.

Let us combine this idea with the product wave function form (2.10). Since we expect physically in the ground state that, in accordance with the Pauli exclusion principle, the two electrons will occupy the same space orbital ψ, with opposed spins, we can write eqn. (2.10) with $\psi_n = \psi_m$. Furthermore, if we could assume that electron 1 feels the full power of the nucleus, then $\psi_{100} = R_{10}(r) Y_{00}(\theta, \phi)$ for the ground-state of a single electron moving in the field of the nucleus of charge Ze would be given by (2.1), except that $\exp(-r/a_0)$ for hydrogen now becomes $\exp(-Zr/a_0)$ for nuclear charge Ze. Since electron 1 say will feel the nuclear attraction reduced somewhat by electron 2, this all suggests that we can take an approximate wave function

$$\Psi(\mathbf{r}_1\mathbf{r}_2) = \exp\left(-\frac{Z'r_1}{a_0}\right) \exp\left(-\frac{Z'r_2}{a_0}\right) \qquad (2.11)$$

where Z' is the 'effective nuclear charge'. On physical grounds it is evident that $Z > Z' > Z-1$, since the value $Z-1$ would obtain only if electron 2 spent all its time nearer to the nucleus than electron 1. This situation is unreasonable physically and therefore Z' should substantially exceed $Z-1$.

Fortunately, there is now a well-defined route for calculating Z' in eqn. (2.11). Thus, we can see from eqn. (2.5) that if we multiply both sides by the complex conjugate wave function Ψ^* (actually, for the ground-state of helium the wave function is real) then we can write immediately

$$E = \frac{\displaystyle\int \Psi^* H \Psi \, d\tau}{\displaystyle\int \Psi^* \Psi \, d\tau}, \qquad (2.12)$$

$d\tau$ indicating integration over all electronic coordinates (for helium $d\tau \equiv d\mathbf{r}_1 \, d\mathbf{r}_2$). We assume the wave function is normalized such that

$$\int \Psi^* \Psi \, d\tau = 1 \qquad (2.13)$$

and then the ground-state energy is given by the numerator of eqn. (2.12). This result, as it stands, is trivial since if Ψ is the exact ground-state wave function, then use of eqns. (2.5) and (2.13) in (2.12) leads to $E = E$!! However, the exact wave function is not known and we must therefore make use of the variational principle of quantum mechanics which says that if we take an approximation Ψ_{trial} (satisfying eqn. (2.13)) to the exact ground-state function Ψ and form the quantity

$$\mathcal{E} = \int \Psi_{\text{trial}}^* \, H \Psi_{\text{trial}} \, d\tau, \qquad (2.14)$$

$d\tau$ again indicating integration over the entire set of coordinates in the wave function, then $\mathcal{E} \geqslant$ the ground-state energy of the Hamiltonian H. Furthermore, knowledge of the trial function to first-order accuracy will allow the energy to be found to higher (second-order) accuracy. This theorem is basic to self-consistent field theories and a proof is therefore given in Appendix 2.2.

We proceed immediately to use the product form (2.11) as Ψ_{trial} in eqn. (2.14), to calculate $\mathcal{E}(Z')$ and then to minimize this latter quantity to find the 'best' effective nuclear charge in the sense that we get as close as possible to the true ground-state energy, with such a trial function.

(a) *Variational calculation for ground state of helium-like system with nuclear charge Ze*

This can now be viewed, if one wishes, as a purely mathematical problem, namely to insert the Hamiltonian (2.6) and the trial function (2.11), when normalized according to eqn. (2.13), into (2.14) for $\mathcal{E}$. However, a little thought enables use to be made of physical arguments and analogies to simplify this calculation. Thus, as discussed above, we write H of eqn. (2.6) in the form

$$H = H_1 + H_2 + \frac{e^2}{r_{12}} \tag{2.15}$$

where

$$H_1 = -\frac{\hbar^2}{2m} \nabla_1^2 - \frac{Ze^2}{r_1} \tag{2.16}$$

is seen to be the Hamiltonian for the motion of electron 1 in the field of the nucleus of charge Ze.

Let us first normalize the trial function, when it is readily found that

$$\Psi_{\text{trial}}(\mathbf{r}_1\mathbf{r}_2) = \frac{Z'^3}{\pi a_0^3} \exp\left(-\frac{Z'}{a_0}\{r_1 + r_2\}\right). \tag{2.17}$$

The terms in $\mathcal{E}$ are evidently:

(i) *Kinetic energy terms*

$-\dfrac{\hbar^2}{2m} \displaystyle\int \Psi \nabla_1^2 \Psi \, d\tau$ and a similar term with ∇_2^2. These are equal, and integrating the term displayed over $\mathbf{r}_2$ we are left with the result

$$-\frac{\hbar^2}{2m} \int \left(\frac{Z'^3}{\pi a_0^3}\right) \exp\left(-\frac{Z'r_1}{a_0}\right) \nabla_1^2\left(\exp\left(-\frac{Z'r_1}{a_0}\right)\right) d\mathbf{r}_1 \tag{2.18}$$

which is evidently the mean kinetic energy $\langle T \rangle$ of an electron moving the field of a nucleus of charge $Z'e$. From the virial theorem in quantum mechanics (see, for example, Slater, 1933; Eyring, Walter and Kimball, 1944), we have for a system in equilibrium under Coulomb forces:

$$2\langle T \rangle + \langle V \rangle = 0 \tag{2.19}$$

where $\langle V \rangle$ is the average potential energy. But evidently

$$\langle T \rangle + \langle V \rangle = E \tag{2.20}$$

and the ground-state energy E for an electron moving in the field of a nucleus of charge $Z'e$ is given by eqn. (2.7) with $n = 1$. Thus we find

$$\langle T \rangle = -E = \frac{Z'^2}{2} \frac{e^2}{a_0} \qquad (2.21)$$

and the kinetic energy terms contribute $Z'^2 e^2/a_0$ to $\mathcal{E}(Z')$.

(ii) *Potential energy terms*

A rather similar argument can be used to evaluate the contribution from the potential energy terms in H_1 and H_2. If the potential term in H_1 had been $-Z'e^2/r_1$, then its average with respect to the trial form Ψ would be simply $\langle V \rangle$, the average potential energy of an electron moving in the field of a nucleus of charge $Z'e$. Then, from eqns. (2.19) and (2.21) $\langle V \rangle$ would be simply $-Z'^2 e^2/a_0$. Then the average of $-Ze^2/r_1$ in eqn. (2.16) with respect to Ψ is $-(Z/Z') \times Z'^2 e^2/a_0$ and thus the two potential energy terms from H_1 and H_2 contribute $-2ZZ'e^2/a_0$. Therefore we can write

$$\mathcal{E}(Z') = \frac{e^2}{a_0}[Z'^2 - 2ZZ'] + \left\langle \frac{e^2}{r_{12}} \right\rangle \qquad (2.22)$$

where explicitly

$$\left\langle \frac{e^2}{r_{12}} \right\rangle = \left(\frac{Z'^3}{\pi a_0^3} \right)^2 e^2 \iint d\mathbf{r}_1 \, d\mathbf{r}_2 \, \frac{e^{-\frac{2Z'r_1}{a_0}} e^{-\frac{2Z'r_2}{a_0}}}{r_{12}} . \qquad (2.23)$$

The nature of the integrand in this equation prompts us to regard it as the electrostatic energy of interaction of two charge distributions $e^{-(2Z'r_1/a_0 d\mathbf{r}_1)}$ and $e^{-(2Z'r_2/a_0 d\mathbf{r}_2)}$. In particular, we can interpret the term

$$\int d\mathbf{r}_2 \, \frac{e^{-\frac{2Z'r_2}{a_0}}}{r_{12}} \qquad (2.24)$$

as the electrostatic potential at $\mathbf{r}_1$ due to a charge distribution of density $e^{-(2Zr_2/a_0)}$. By a simple change of the unit of length, we see that we need the electrostatic potential, χ say, due to a charge density e^{-r}, which we can evidently obtain by solving the Poisson equation

3*

$$\nabla^2\chi = -4\pi e^{-r}. \tag{2.25}$$

χ will be spherically symmetrical for the spherical charge distribution under discussion and therefore we need only the radial part of the Laplacian ∇^2 in eqn. (1.8). Rewriting $r^{-2}(\partial/\partial r)(r^2(\partial\chi/\partial r))$ in the equivalent form $r^{-1}(d^2/dr^2)(r\chi)$ we have

$$\frac{d^2}{dr^2}(r\chi) = -4\pi r e^{-r} \tag{2.26}$$

and integrating both sides with respect to r we obtain

$$\frac{d}{dr}(r\chi) = -4\pi\int re^{-r}\,dr$$

$$= 4\pi r e^{-r} + 4\pi e^{-r} + k_1 \tag{2.27}$$

where k_1 is the first of two constants of integration. A further integration yields in a similar way

$$\chi(r) = -\frac{8\pi}{r}e^{-r} - 4\pi e^{-r} + k_1 + \frac{k_2}{r}. \tag{2.28}$$

As $r \to \infty$, $\chi \to 0$ and therefore k_1 is zero. Furthermore, eqn. (2.28) gives the potential due to a volume distribution of charge and therefore there is no $1/r$ singularity in the potential as $r \to 0$, for this would herald a point charge at the origin. Hence $k_2 = 8\pi$ and we have the final result

$$\chi(r) = \frac{4\pi}{r}[2 - 2e^{-r} - re^{-r}]. \tag{2.29}$$

Using this result to evaluate the integral (2.24) and inserting this value into (2.23) we find, writing $s = 2Z'r/a_0$,

$$\left\langle\frac{e^2}{r_{12}}\right\rangle = \left(\frac{Z'^3}{\pi a_0^3}\right)^2 e^2 \int ds_1\,ds_2\,\frac{e^{-s_1}e^{-s_2}}{s_{12}}\left(\frac{a_0}{2Z'}\right)^5$$

$$= \frac{Z'}{32\pi^2}\frac{e^2}{a_0}\int ds_1\chi(s_1)\,e^{-s_1}$$

$$= \frac{Z'e^2}{32\pi^2 a_0}\int_0^\infty ds_1 4\pi s_1^2\chi(s_1)\,e^{-s_1}. \tag{2.30}$$

The integral is readily evaluated by inserting (2.29) into the above equation, and has the value $20\pi^2$. Thus, from eqns. (2.22) and (2.30) it follows that

$$\mathscr{E}(Z') = \frac{e^2}{a_0} \left[Z'^2 - 2ZZ' + \frac{5}{8} Z' \right] \qquad (2.31)$$

and hence, to find the minimum of $\mathscr{E}(Z')$, which will bring $\mathscr{E}$ as near as possible to the ground-state energy E, we must have

$$0 = \frac{d\mathscr{E}}{dZ'} = \frac{e^2}{a_0} \left[2Z' - 2Z + \frac{5}{8} \right] \qquad (2.32)$$

or

$$Z' = Z - \tfrac{5}{16}. \qquad (2.33)$$

As anticipated, the effective nuclear charge lies between Z and $Z-1$. The corresponding approximation to the energy obtained by inserting $Z' = Z - \tfrac{5}{16}$ in eqn. (2.31) is simply

$$\mathscr{E}_{min} = -\left(Z - \frac{5}{16}\right)^2 \frac{e^2}{a_0}. \qquad (2.34)$$

For the helium atom, with $Z = 2$, we find

$$\mathscr{E}_{min} = -2{\cdot}85 \text{ au} \qquad (2.35)$$

which is in pretty good agreement with the measured value of -2.90 au.

In passing we note that first-order perturbation theory (see any book on quantum mechanics) would calculate the average of e^2/r_{12} with respect to the 'unperturbed' wave function with $Z' = Z$. Evidently, from eqn. (2.31) this average is $\frac{5}{8}(Ze^2/a_0)$ and for He we correct the unperturbed energy -4 au by this value of $1{\cdot}25$ au, to obtain $-2{\cdot}75$ au. The value given in eqn. (2.35) is lower than this, as it must be from the variational principle.

We have carried through this calculation in detail as it illustrates some of the points we want to emphasize in this book. Thus, we have seen how to use the variation principle to approximate to one-electron wave functions for atoms, and also the way in which, in He, the second electron must affect the field in which the first electron moves.

We shall now see how the variational calculation outlined above can be extended in a powerful way, provided only that we restrict ourselves to the product form (2.10) for Ψ.

(b) *Best possible orbitals for helium atom*

Suppose we start out, not from the product (2.11) but from a general form

$$\Psi(\mathbf{r}_1\mathbf{r}_2) = \psi(\mathbf{r}_1)\,\psi(\mathbf{r}_2) = P(r_1)\,P(r_2)/r_1 r_2 \qquad (2.36)$$

since $P = r\psi$ is a convenient quantity to work with.

We shall assume, for convenience, that

$$\int_0^\infty P^2\, dr = 1. \qquad (2.37)$$

The variational energy is again given by eqn. (2.12) with the trial function now defined by eqn. (2.36). Following through the argument as we did with the 'effective nuclear charge' $Z-\frac{5}{16}$, we can write the energy (compare Hartree, 1957) as[†]

$$\mathcal{E} = 2I + F_0 \qquad (2.38)$$

where

$$F_0 = \int \frac{P^2(r_1)}{4\pi r_1^2} \int \frac{1}{r_{12}} \frac{P^2(r_2)}{4\pi r_2^2}\, d\mathbf{r}_1\, d\mathbf{r}_2 \qquad (2.39)$$

which is again the mutual potential energy of two spherically symmetric distributions of charge, with density

$$\varrho(r) = P^2(r)/4\pi r^2, \qquad (2.40)$$

in exact analogy with the previous argument used to evaluate $\langle e^2/r_{12}\rangle$ in eqn. (2.23).

In terms of the electrostatic potential $\chi(r)$ due to this charge distribution, we can write, with

$$\chi(r) = Y(r)/r \qquad (2.41)$$

[†] Here we follow Hartree in using atomic units in which $e = m = \hbar = 1$. The unit of length is then $a_0 = \hbar^2/me^2$ and the unit of energy e^2/a_0 (cf. p. 21).

$$F_0 = \int_0^\infty P^2(r) \frac{1}{r} Y(r) \, dr. \tag{2.42}$$

We also have, from the other terms than the electron–electron interaction

$$I = -\frac{1}{2} \int_0^\infty P(r) \left[P''(r) + \frac{4}{r} P(r) \right] dr. \tag{2.43}$$

Hence, we have the variational energy $\mathcal{E}$ in terms of $P(r)$ and $Y(r)$.

(i) *Minimization with respect to $P(r)$*

Obviously, if we make no restrictions on $P(r)$ except normalization, we can express the changes ΔI and ΔF_0 in terms of a general variation $\Delta P(r)$ of $P(r)$.

The first order change in I takes the form

$$\Delta I = -\frac{1}{2} \int_0^\infty \Delta P(r) \left[\frac{d^2}{dr^2} + \frac{4}{r} \right] P(r) \, dr$$

$$-\frac{1}{2} \int_0^\infty P(r) \left[\frac{d^2}{dr^2} + \frac{4}{r} \right] \Delta P(r) \, dr. \tag{2.44}$$

The contributions are readily shown to be the same and hence we can write

$$\Delta I = -\int_0^\infty \Delta P(r) \left[\frac{d^2}{dr^2} + \frac{4}{r} \right] P(r) \, dr. \tag{2.45}$$

Turning to the electron-electron energy, we find

$$\Delta F_0 = \frac{1}{(4\pi)^2} \int \frac{1}{r_{12}} [2P(r_1) \, \Delta P(r_1) \, P^2(r_2)$$

$$+ P^2(r_1) \, 2P(r_2) \, \Delta P(r_2)] \, dr_1 \, dr_2 \, d\omega_1 \, d\omega_2 \tag{2.46}$$

which, on performing the angular integrations over ω_1 and ω_2 yields

$$\Delta F_0 = 4 \int_0^\infty \Delta P(r) \frac{1}{r} Y(r) P(r) \, dr. \tag{2.47}$$

Hence, the variation $\Delta \mathcal{E}$ is given by

$$\Delta \mathcal{E} = 2 \, \Delta I + \Delta F_0$$

$$= -2 \int_0^\infty \Delta P(r) \left[\frac{d^2}{dr^2} + \frac{4}{r} - \frac{2Y(r)}{r} \right] P(r) \, dr. \qquad (2.48)$$

But we must add a Lagrange multiplier to take care of the normalization (2.37) and consider

$$\mathcal{E}' = \mathcal{E} + \lambda \int_0^\infty P^2 \, dr. \qquad (2.49)$$

We find then the desired Schrödinger equation for $P = r\psi$

$$\left[\frac{d^2}{dr^2} + \frac{4}{r} - \frac{2Y}{r} - \lambda \right] P = 0, \qquad (2.50)$$

with $d^2 Y/dr^2 = -P^2/r$, which is simply Poisson's equation.

2.3. Hartree's self-consistent field

The message in eqn. (2.50), which determines the 'best' orbital $\psi = P/r$ for the helium atom is now clear. On to the potential energy of the nucleus ($-2/r$ in the units used in eqn. (2.50)) is added a term Y/r, where this term is taking account simply of the electrostatic potential created by the charge cloud of the other electron, the density being $\psi^2 = P^2/r^2$. But, of course, we do not know the wave function ψ at the outset. Therefore, we must guess the term Y/r initially, integrate equation (2.50) to find P and hence a new $Y(r)$. With this new field, we re-solve the Schrödinger eqn. (2.50), and we continue until a given $Y(r)$ reproduces itself to our specified numerical accuracy. We then have established the *self-consistent field*[†] in which an electron moves. Clearly, we could use as a starting point in solving eqn. (2.50), the approximate ψ we obtained earlier, namely $\psi = e^{-(Z - \frac{5}{16})r}$, where $Z = 2$ for He. We saw explicitly above how to calculate the electrostatic potential due to a charge distribution of this form (cf. eqn. (2.29)). We

[†] For a fuller discussion, see reprint 1 by Hartree on p. 167 of this volume.

would then add this to the electron-nucleus potential energy $-2e^2/r$ to obtain a first estimate of the self-consistent field. But in fact the self-consistent solution of eqn. (2.50) is available to us, from the numerical work of Hartree himself, and particularly from the investigation of Wilson and Lindsay (1935).

Let us now briefly discuss the more general case when we have Z electrons, with $Z > 2$. We clearly have a much more complicated Schrödinger equation, with a kinetic energy term $(-\hbar^2/2m) \nabla^2$ in the Hamiltonian for each of the Z electrons, and furthermore terms e^2/r_{ij} in the potential energy, representing the Coulombic repulsion between the ith and jth electrons. Strictly we must solve for the total wave function Ψ of the atom, which will depend on the coordinates $r_1, r_2, \ldots r_Z$ of all the Z electrons and also on their spin coordinates $\sigma_1 \ldots \sigma_Z$. Such a wave function, needless to say, becomes extremely complex for a heavy atom such as Hg $(Z = 80)$ or U $(Z = 92)$.

Fortunately, it was Hartree (1927; see reprint 1 on. p. 167 of this volume), who pointed out that because of the essential dominance of the nuclear attraction in an atomic problem, it was for any atom, no matter with how few or how many electrons, a useful first approximation to ascribe to each electron its own individual wave function ϕ_i and its own individual energy E_i. Then supposing we assume that all electrons move in the same potential energy $V(r)$, we can make use of the separation of the wave function ϕ_i as given by eqn. (1.9), and we can reduce the problem to finding the radial wave functions R_{nl} which satisfy eqn. (1.13). As remarked earlier, this is a simple matter on an electronic computer, for a given choice of $V(r)$, and we shall consider such solutions for specific potentials later.

But first let us go straight to the heart of the Hartree self-consistent field concept: the determination of $V(r)$. We have dealt with this explicitly for helium above: now we treat the more general case $Z > 2$.

(a) *Self-consistent field in heavy atoms*

Suppose we have solved for the lowest energy levels E_i and their corresponding wave functions ϕ_i. Then, since we know that an electron described by ϕ_i has a probability density ϕ_i^2 (assuming ϕ_i is real: otherwise $|\phi_i|^2$ must be used), we can say immediately that the electron density $\varrho(\mathbf{r})$ in the atom is given by

$$\varrho(\mathbf{r}) = \sum_{\substack{\text{all occupied} \\ \text{levels}}} \phi_i^2 \qquad (2.51)$$

But it is extremely plausible to argue that the potential energy in which the electrons move is that due first to the nucleus, and given by $-Ze^2/r_i$ for the ith electron at distance r_i from the nucleus, and then a contribution from the electronic charge density given by eqn. (2.51). This contribution $V_e(r)$ say, is clearly obtained by solving the Poisson equation of electrostatics

$$\nabla^2 V_e(r) = 4\pi e \varrho(r). \qquad (2.52)$$

But from the Schrödinger equation, for an assumed total potential energy $V(r)$, we can find the wave functions ϕ_i, and hence we can calculate $\varrho(r)$ from eqn. (2.51). Solving eqn. (2.52), we shall find in general that the potential $-Ze^2/r + V_e(r)$ does not exactly reproduce the starting potential. The calculation is not self-consistent and we must modify our original guess. We have only established the self-consistent field when the wave functions calculated from the assumed potential field reproduce the original potential to some specified accuracy.

This idea has proved extremely fruitful in many branches of physics concerned with interacting particles. It requires generalizing in various directions, but the essential concept is already apparent from the discussion above. Three points that need further consideration are:

(i) An electron does not act on itself, and so the assumption of a common potential for each electron is not quite correct.

(ii) The 'best possible' choice of potential is not as simple as that discussed above, for another, and somewhat more profound, reason than (i). The best potential depends on velocity or energy, and not simply on position $\mathbf{r}$.

(iii) We have not specified how to build an approximate total wave function Ψ from the individual wave functions ϕ_i, except for He, for which eqns. (2.36) or (2.17) solve this problem.

It will be clear from our discussion of the self-consistent field for helium that for Z electrons we will rely again on the variation principle, combined with a product trial wave function. Such a calculation is carried out in Appendix 2.3, and leads to the Hartree equations generalizing the result (2.50) we obtained for helium. For the single-particle orbital ϕ_i of the ith electron we must now solve [cf. eqn. (A2.3.5)]

$$\left[-\frac{\hbar^2}{2m} \nabla_i^2 - \frac{Ze^2}{r_i} + e^2 \sum_{j \neq i} \int \frac{|\phi_j(r_j)|^2}{r_{ij}} \, d\mathbf{r}_j \right] \phi_i(r_i) = \varepsilon_i \phi_i(r_i). \quad (2.53)$$

The physical interpretation of these equations should by now be perfectly clear. The ith electron on which we have focused attention moves in the average field created by the other $N-1$ electrons and the nucleus. Thus, point (i) referred to above in the description of the common self-consistent field in which each electron moves in the same potential is taken care of in the Hartree theory. The eigenvalue ε_i entered the theory as a Lagrange multiplier to take care of normalization. Though the total energy of the atom or ion is related to $\sum_i \varepsilon_i$, the summation being over all occupied energy levels, further investigation shows that this formula has to be corrected because it counts electron–electron interactions twice over.

Equations (2.53) have been solved for many atoms and ions, much of the pioneering work being done by D. R. Hartree and his father W. Hartree. Unfortunately, the results of many such calculations have had to be obtained by purely numerical methods and the results presented in extensive tables (see, for example, Hartree, 1957; Herman and Skillman, 1963). It is not our intention here to discuss such numerical procedures, but rather to try to expose some of the physical results which emerge. To do this, we shall often be content with somewhat simplified arguments, knowing that we can check and, if necessary, refine the results, by appeal to the numerical solutions. With such an aim in mind, we consider next an approximate formulation of the Hartree method for atoms, which, while rough in some respects, supplies an interesting and universal description of atomic structure.

Problems

1. Starting from Newton's equations

$$m\ddot{x} = F_x$$

for a particle of mass m acted on by a force $\mathbf{F} = (F_x, F_y, F_z)$ and using the identity

$$\frac{d}{dt}[x\dot{x}] = x\ddot{x} + \dot{x}^2,$$

prove the classical virial theorem

$$2\overline{T} + \overline{\mathbf{r} \cdot \mathbf{F}} = 0$$

where T is the kinetic energy, $\mathbf{r} = (x, y, z)$ and the bars denote time averages. (This is to be contrasted with quantum-mechanical form, in which averages are the usual expectation values of wave mechanics.)

2. Calculate the quantum mechanical average of r^2 for the electron in the hydrogen atom for (a) the *1s* ground state and (b) the *2s* state of the degenerate first excited level. (*N.B.* We shall see in Chapter 5 that $\langle r^2 \rangle$ determines the diamagnetic susceptibility of electrons in atoms.)

How will $\langle r^2 \rangle$ vary with Z in the ground-state of a hydrogen-like ion of nuclear charge Ze?

CHAPTER 3

Thomas–Fermi atom

WHAT we require is a means of solving the Schrödinger eqn. (1.1) for a general potential energy $V(\mathbf{r})$, so that we can construct the electron density $\varrho(\mathbf{r})$, and hence employ self-consistency to determine $V(\mathbf{r})$.

A rough way of relating $\varrho(\mathbf{r})$ and $V(\mathbf{r})$ is to utilize a semi-classical approximation due to Thomas (1926; see reprint 2 at the end of the volume) and independently discovered by Fermi (1928, reprint 3). We turn straight away to the physical basis of the method, and shall only later endeavour to display its connection directly with the wave equation.

3.1. Classical energy equation for fastest electron

Suppose that the self-consistent potential energy $V(\mathbf{r})$ is given. Then the fastest electron, with total energy E_f say, may be described by the classical energy equation

$$E_f = \frac{p_f^2}{2m} + V(\mathbf{r}). \tag{3.1}$$

Here m is the electron mass and p_f is clearly the maximum momentum of the electron and if the maximum energy E_f is constant, as it must be, for otherwise electrons could redistribute themselves to lower the energy, then p_f must vary in space, since $V(\mathbf{r})$ depends on position. Hence we may write

$$p_f^2(\mathbf{r}) = 2m[E_f - V(\mathbf{r})]. \tag{3.2}$$

If we can now relate the maximum momentum to the electron density, then we shall have achieved the desired relation between $\varrho(\mathbf{r})$ and V.

3.2. Electron density and maximum electronic momentum

This can be done in an approximate way as follows. Consider the electrons around position $\mathbf{r}$. In momentum space, these electrons fill a sphere of radius $p_f(\mathbf{r})$. If we now consider unit volume of coordinate space, the volume of occupied phase space (product of volumes of coordinate space and momentum space) is $\frac{4}{3}\pi p_f^3(\mathbf{r})$. But if we wish to retain the classical description in terms of the phase space, the Heisenberg Uncertainty Principle restricts us to deal with cells of volume h^3. These cells correspond to electron energy levels and can hold two electrons each, provided the electron spins are opposed. In this manner, we can build up a semi-classical theory which includes both the Heisenberg Relation and the Pauli Exclusion Principle. It then follows immediately that the electron density $\varrho(\mathbf{r})$ is given by

$$\varrho(\mathbf{r}) = \frac{2}{h^3} \cdot \frac{4\pi}{3}\, p_f^3(\mathbf{r}). \tag{3.3}$$

Hence, combining eqns. (3.2) and (3.3) we find the Thomas–Fermi relation between density ϱ and potential V:

$$\varrho(\mathbf{r}) = \frac{8\pi}{3h^3}\,(2m)^{\frac{3}{2}}\,[E_f - V(\mathbf{r})]^{\frac{3}{2}}. \tag{3.4}$$

This relation purports essentially to describe the result of solving for the one-electron wave functions ϕ_i in the self-consistent field, and then forming the sum of the squares of these wave functions over the occupied levels, as in eqn. (2.51).

Clearly, this result (3.4) cannot be exact. For when $V(\mathbf{r}) > E_f$, we cannot use eqn. (3.4) and the best we can do in this theory is then to put $\varrho = 0$ in this classically forbidden region. This is in contrast to the situation we know to obtain wave mechanically, where electron wave functions can leak through to the classically forbidden region and have non-zero amplitude there.

We should also comment at this point that the maximum energy E_f is to be determined from the normalization condition that $\varrho(\mathbf{r})$ integrated through all space must give the total number of electrons N in the atom or ion, that is

$$\int \varrho(\mathbf{r}) \, d\mathbf{r} = N. \tag{3.5}$$

Evidently for the neutral atom we must have $N = Z$, the atomic number.

3.3. Electron density for Coulomb field

Though in a real atom $V(r)$ must be determined self-consistently, let us immediately illustrate this approximation by applying the theory to a pure Coulomb field. Then we have for $r < Ze^2/|E_f|$

$$\varrho(\mathbf{r}) = \frac{8\pi}{3h^3} (2m)^{\frac{3}{2}} \left[E_f + \frac{Ze^2}{r} \right]^{\frac{3}{2}} \tag{3.6}$$

and if we take the case of Z electrons, as in a neutral atom, eqn. (3.6) is readily integrated over $\mathbf{r}$, leading to E_f as

$$-E_f = \frac{Z^{\frac{4}{3}} e^2}{a_0} \frac{1}{18^{\frac{1}{3}}}. \tag{3.7}$$

As we have already remarked the approximations used are objectionable in the sense that when

$$\frac{Ze^2}{r} + E_f < 0 \tag{3.8}$$

the electron density must vanish. Furthermore, as we approach the nucleus, $\varrho(r) \sim r^{-\frac{3}{2}}$, whereas the only wave functions which are non-zero at the nucleus, the s states, are certainly finite, as seen explicitly from eqns. (2.1) and (2.2). We must be very careful, therefore, to remember these limitations in applying the method. However, let us proceed immediately to study the general features of the self-consistent field in atoms in this approximation.

3.4. Approximate self-consistent fields in atoms and ions

We must now combine eqn. (3.4) with Poisson's equation

$$-\nabla^2 V = 4\pi e^2 \varrho(\mathbf{r}) \qquad (3.9)$$

and if we write

$$V(\mathbf{r}) - E_f = -\frac{Ze^2}{r} \phi \qquad (3.10)$$

and

$$r = bx \qquad (3.11)$$

then ϕ is immediately dimensionless and, if b is chosen suitably, then as shown below, x is a convenient dimensionless measure of distance from the nucleus. We then find the so-called dimensionless Thomas–Fermi equation:

$$\frac{d^2\phi}{dx^2} = \frac{\phi^{\frac{3}{2}}}{x^{\frac{1}{2}}} \qquad (3.12)$$

this form arising, with the coefficient of $\phi^{\frac{3}{2}} \big/ x^{\frac{1}{2}}$ as unity on the right-hand side, from the choice

$$b = \frac{1}{4} \left(\frac{9\pi^2}{2Z} \right)^{\frac{1}{3}} a_0 = \frac{0 \cdot 8853 a_0}{Z^{\frac{1}{3}}}. \qquad (3.13)$$

Thus, we have established the 'scale' of the Thomas–Fermi atom model, since it can be seen that b is a length, which also decreases with increasing Z according to eqn. (3.13).

Since we are assuming that V is the total potential including that of the nucleus, it is clear from eqn. (3.10) that, as $r \to 0$, we must have

$\phi(0) = 1$ since $V(r) \to -Ze^2/r$ as we approach the nucleus having charge Ze. The boundary condition $\phi(0) = 1$ holds, of course, both for neutral atoms and for positive ions (negative ions are not stable, it turns out, in the Thomas–Fermi theory). We shall shortly discuss the further boundary condition needed in order to solve the second-order differential eqn. (3.12) uniquely.

Before doing so, we record here that an *exact* solution of (3.12) is readily shown to be

$$\phi(x) = \frac{144}{x^3} \tag{3.14}$$

but this does not satisfy the boundary condition $\phi(0) = 1$.

(a) *Physical interpretation of positive ions in Thomas–Fermi theory*

To discuss the physical interpretation of the various solutions of eqn. (3.12) we start by differentiating eqn. (3.10) with respect to r when we find

$$\frac{dV}{dr} = \frac{Ze^2}{r^2}\phi - \frac{Ze^2}{r}\frac{\partial\phi}{\partial r}. \tag{3.15}$$

However, V can be written as $-e \times$(electrostatic potential) and since the electric field $\mathbf{E} = -\text{grad}$ (electrostatic potential) we can write

$$|\mathbf{E}| = \frac{1}{e}\frac{\partial V}{\partial r} = \frac{Ze}{r^2}\phi - \frac{Ze}{r}\frac{\partial\phi}{\partial r}. \tag{3.16}$$

The solutions of eqn. (3.12) with $\phi(x_0) = 0$ for finite x_0 evidently correspond to zero electron density outside a radius $R_i = bx_0$, and let us suppose that such an electron cloud contains N electrons. Then we may write, for $r \geqslant R_i$,

$$|\mathbf{E}| = \frac{1}{e}\frac{\partial V}{\partial r} = \frac{1}{e}\frac{d}{dr}\left\{-\frac{(Z-N)e^2}{r}\right\}. \tag{3.17}$$

Now we equate these two expressions (3.16) and (3.17) for $|\mathbf{E}|$ at $r = R_i$ $= bx_0$ and since $\phi(x_0) = 0$ we find

$$
\left.
\begin{aligned}
-\frac{Ze}{b^2 x_0}\left(\frac{\partial\phi}{\partial x}\right)_{x_0} &= \frac{(Z-N)e}{b^2 x_0^2} \\[2mm]
-x_0\left(\frac{\partial\phi}{\partial x}\right)_{x_0} &= \left(1-\frac{N}{Z}\right).
\end{aligned}
\right\}
\tag{3.18}
$$

or

It is clear from Fig. 3.1, which shows the nature of the solutions of eqn. (3.12) with $\phi(0) = 1$, that $(\partial\phi/\partial x)_{x_0}$ is negative, and hence from eqn. (3.18) those solutions with $\phi(x_0) = 0$ for finite x_0 correspond to $N/Z < 1$, that is to *positive ions*. The elementary construction shown in Fig. 3.1 indicates how we obtain the ionic charge, given such a solution $\phi(x)$ of the dimensionless Thomas–Fermi equation (3.12).

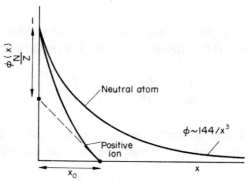

FIG. 3.1. Physical interpretation of solutions of the dimensionless Thomas–Fermi equation (3.12).

The special case $N/Z = 1$, i.e. the neutral atom, evidently corresponds to

$$
-x_0\left(\frac{\partial\phi}{\partial x}\right)_{x_0} = 0
\tag{3.19}
$$

and this is seen after a little thought to be the case $x_0 \to \infty$. This is verified by noting that the solution $\phi(x)$ which tends to zero at infinity has the form (3.14) for sufficiently large x. Clearly we have then $(x\,\partial\phi/\partial x) \sim -3\cdot144/x^3$ for large x, which tends to zero as $x \to \infty$.[†]

The dimensionless Thomas–Fermi eqn. (3.12) has been solved numerically for a variety of values of x_0 and solutions for three finite values of x_0 as well as for the isolated atom with $x_0 = \infty$ are recorded in Appendix 3.1.

We shall now proceed to discuss how the binding energies of atoms and ions can be related to these solutions of the dimensionless Thomas–Fermi equation.

Problems

1. For solutions of the dimensionless Thomas–Fermi eqn. (3.12) which satisfy the condition $\phi(0) = 1$, show that there is an expansion for small x of the form

$$\phi(x) = 1 + a_2 x + a_3 x^{\frac{3}{2}} + a_4 x^2 + \dots$$

Determine a_3 and a_4. (This expansion is due to Baker (1930). Note that it will generate, for small x, the various solutions shown schematically in Fig. 3.1, as the initial slope $a_2 = \phi'(0)$ is varied.)

2. Express the average value of r^2 for an electron in the Thomas–Fermi atom in terms of the dimensionless solution $\phi(x)$ satisfying the conditions $\phi(0) = 1$ and $\phi \to 0$ at infinity.

3. Using the Thomas–Fermi model for a Coulomb field as set out in section 3.3, find the energy below which just two electrons are to be found. Compare the total energy of these two electrons with the exact value for non-interacting electrons in a Coulomb field.

[†] Since $V \to 0$ as $r \to \infty$, and $\phi \to 0$ as $r(\alpha x) \to \infty$, $E_f = 0$ for the self-consistent Thomas–Fermi atom, in contrast to the Coulomb field case of section 3.3.

CHAPTER 4

Energies of atoms and ions

IN this chapter we shall consider the total binding energies of atoms and ions. Because the theory is simpler than the Hartree theory, and can be put into universal form applicable to a general ion of nuclear charge Ze and with N electrons, we begin the discussion with the Thomas–Fermi atom, discussed in the previous chapter.

There we obtained the density-potential relation (3.4) characteristic of that theory, by applying free electron relations locally.[†] We now use the same arguments to set up the total energy in the Thomas–Fermi approximation.

4.1. Variational derivation of Thomas–Fermi equation

The total energy E in this approximation is made up of two parts; the kinetic energy T of the electronic charge cloud and the total potential energy V which will also consist of two parts

$$V = V_{en} + V_{ee} \tag{4.1}$$

where V_{en} is the potential energy of the electron–nucleus interaction while V_{ee} is the electrostatic interaction energy of the charge cloud with density $\varrho(\mathbf{r})$. Thus we have explicitly

$$V = \int \varrho(\mathbf{r}) V_N \, d\mathbf{r} + \frac{1}{2} e^2 \int \frac{\varrho(\mathbf{r}) \, \varrho(\mathbf{r}')}{|\mathbf{r} - \mathbf{r}'|} \, d\mathbf{r} \, d\mathbf{r}' , \tag{4.2}$$

[†] The reader will recognize readily that eqns. (3.3) and (3.4) become *exact* for a uniform electron gas.

V_N being the electron–nuclear interaction, which in an atom or ion with charge Ze on the nucleus is simply $(-Ze^2/r)$.

To set up the kinetic energy T, we note that, in the free electron theory, the mean kinetic energy of N electrons in a volume ϑ is given by

$$\text{Total K.E.} = N \int_0^{p_f} \frac{p^2}{2m} \frac{4\pi p^2 \, dp}{\frac{4}{3}\pi p_f^3} \qquad (4.3)$$

the factor $p^2/2m$ being simply the kinetic energy of an electron of momentum p while $4\pi p^2 \, dp / \frac{4}{3}\pi p_f^3$ is the probability of an electron in the Fermi gas having momentum between p and $p+dp$. Thus we find for the free electron gas

$$\text{Total K.E.} = N \frac{3}{2mp_f^3} \frac{p_f^5}{5} = \frac{3}{5} N \frac{p_f^2}{2m} \qquad (4.4)$$

and hence the mean kinetic energy per particle is $\frac{3}{5}$ of the Ferm energy $p_f^2/2m = E_f$.

Thus the kinetic energy per unit volume is $\frac{3}{5}(N/\vartheta)(p_f^2/2m)$ and making use of eqn. (3.3) for free electrons we have

$$\text{Kinetic energy density} = \frac{3}{10m}\left(\frac{N}{\vartheta}\right)\left(\frac{3h^3}{8\pi}\right)^{\frac{2}{3}}\left(\frac{N}{\vartheta}\right)^{\frac{2}{3}}$$

$$= c_k \left(\frac{N}{\vartheta}\right)^{\frac{5}{3}} : c_k = \frac{3h^2}{10m}\left(\frac{3}{8\pi}\right)^{\frac{2}{3}}. \qquad (4.5)$$

Using the relation (4.5) locally, $(N/\vartheta) \to \varrho(\mathbf{r})$ and the kinetic energy density $\alpha\{\varrho(\mathbf{r})\}^{\frac{5}{3}}$, yielding

$$T = c_k \int \varrho^{\frac{5}{3}} \, d\mathbf{r}. \qquad (4.6)$$

Hence, from eqns. (4.6) and (4.2) we find the total energy E as

$$E = c_k \int \varrho^{\frac{5}{3}} \, d\mathbf{r} + \int \varrho(\mathbf{r})V_N \, d\mathbf{r} + \frac{1}{2}e^2 \int\int \frac{\varrho(\mathbf{r})\,\varrho(\mathbf{r}')}{|\mathbf{r}-\mathbf{r}'|} \, d\mathbf{r}\,d\mathbf{r}'. \qquad (4.7)$$

We are now in a position to evaluate E from the solutions of the self-consistent Thomas–Fermi equation and we shall do this below.

But before that, we establish what is one of the pillars of this approximate theory, namely that the Thomas–Fermi equation follows from eqn. (4.7) by minimizing E with respect to variations in the density ϱ. Of course, in carrying out such a variation, we must always ensure that the number of electrons remains fixed at the given number N in the ion ($N = Z$ for a neutral atom, of course). Thus we require to make $\delta E = 0$ for arbitrary small variations $\partial \varrho$ in ϱ, subject only to the normalization condition

$$\int \varrho(\mathbf{r})\, d\mathbf{r} = N. \tag{4.8}$$

The way to deal with this subsidiary condition is, as we have seen earlier, via a Lagrange multiplier, λ say, when we can write the variation principle as

$$\delta(E + \lambda N) = 0. \tag{4.9}$$

It is now straightforward from eqn. (4.7) to calculate the variation δE in E due to a change $\delta \varrho$ in ϱ. We then find immediately

$$\delta E = c_k \int \frac{5}{3} \varrho^{\frac{2}{3}}\, \delta \varrho\, d\mathbf{r} + \int \delta \varrho V_N\, d\mathbf{r} + \frac{1}{2}\, e^2 \int \frac{\delta \varrho(\mathbf{r})\, \varrho(\mathbf{r}')}{|\mathbf{r}-\mathbf{r}'|}\, d\mathbf{r}\, d\mathbf{r}'$$

$$+ \frac{1}{2}\, e^2 \int \frac{\varrho(\mathbf{r})\, \delta \varrho(\mathbf{r}')}{|\mathbf{r}-\mathbf{r}'|}\, d\mathbf{r}\, d\mathbf{r}'. \tag{4.10}$$

From symmetry considerations, the last two terms on the right-hand side of eqn. (4.10) are equal and since the electrostatic potential $\chi_e(\mathbf{r})$ of the electronic cloud of charge density $-e\varrho(r)$ is given by

$$\chi_e(\mathbf{r}) = -e \int \frac{\varrho(\mathbf{r}')}{|\mathbf{r}-\mathbf{r}'|}\, d\mathbf{r}', \tag{4.11}$$

we can write, putting the potential energy of an electron sitting in the electrostatic potential $\chi_e(\mathbf{r})$ as $V_e(\mathbf{r}) \equiv (-e\chi_e(\mathbf{r}))$

$$0 = \delta(E + \lambda N) = \int \delta \varrho \left[\tfrac{5}{3} c_k \varrho^{\frac{2}{3}} + V_N(\mathbf{r}) + V_e(\mathbf{r}) + \lambda \right] d\mathbf{r} \tag{4.12}$$

and for this to be true for arbitrary variations $\delta\varrho$ we must have

$$\tfrac{5}{3} c_k \varrho^{\frac{2}{3}} = -[\lambda + V(\mathbf{r})] : V(\mathbf{r}) = V_N + V_e. \qquad (4.13)$$

But this is immediately seen to be equivalent to eqn. (3.4), the Fermi energy E_f being identified with the Lagrange multiplier through $E_f \equiv -\lambda$. We deal a little more basically with such an identification below. Thus, in more technical language, we can say that the Thomas–Fermi relation (3.4) is the Euler equation of the variation problem (4.9), with E given by eqn. (4.7). We stress that this variational principle is *not* the same as that for the ground-state energy in terms of a trial wave function. In particular, with approximations of a semi-classical kind such as underly the Thomas–Fermi theory, we have no assurance that the energy we calculate will be an upper bound to the true ground state energy. In practice, the Thomas–Fermi atom has a *lower* energy than the true ground state.

To deal a little more fully with the identification of the Lagrange multiplier $(-\lambda)$ with the Fermi energy, we can note that the total energy E is a function Z and N, and we shall obtain an explicit expression for $E(Z, N)$ below in the Thomas–Fermi approximation.

To compare with the variation principle (4.9), we may also write

$$\delta E = \frac{\partial E}{\partial N}\, \delta N \qquad (4.14)$$

and hence we see that

$$\frac{\partial E}{\partial N} = -\lambda. \qquad (4.15)$$

But from general thermodynamic considerations we have

$$\theta\, dS = dE + p\, d\mathcal{O} - \mu\, dN \qquad (4.16)$$

where S is the entropy, θ the absolute temperature, E the internal energy, p the pressure and $\mathcal{O}$ the volume. The chemical potential μ is evidently thereby such that $E_f = -\lambda$ is this chemical potential at $T = 0$.

4.2. Evaluation of total electronic energy for heavy ions

For a system in equilibrium under Coulomb forces we have the virial theorem (2.19), and hence the total electronic energy E is given by

$$E = -T \tag{4.17}$$

where T is the quantum-mechanical average of the kinetic energy.

In the Thomas–Fermi theory, we have to evaluate the kinetic energy T via eqn. (4.6) from a knowledge of $\varrho(\mathbf{r})$, the number of electrons per unit volume, related to the total potential energy $V(\mathbf{r})$ by eqn. (3.4).

To evaluate eqn. (4.6) for the self-consistent Thomas–Fermi theory by direct manipulation is possible, the rather lengthy derivation being due, essentially, to Milne (1927). We show below that

$$E = -T = \frac{12}{7} \left(\frac{2}{9\pi^2}\right)^{\frac{1}{3}} \left[\phi'(0) + \left(1 - \frac{N}{Z}\right)^2 \frac{1}{x_0}\right] Z^{\frac{7}{3}} \frac{e^2}{a_0}$$

$$= 0.4841 \left[\phi'(0) + \left(1 - \frac{N}{Z}\right)^2 \frac{1}{x_0}\right] Z^{\frac{7}{3}} \frac{e^2}{a_0}. \tag{4.18}$$

It should be noted that when $N = Z$, we know from numerical solution of the dimensionless Thomas–Fermi equation (see Appendix 3.1) that

$$\phi'(0) = -1.58805; \qquad x_0 = \infty \tag{4.19}$$

and we obtain the Thomas–Fermi result for atomic binding energies

$$E(Z, Z) = -0.7687 Z^{\frac{7}{3}} \frac{e^2}{a_0}. \tag{4.20}$$

This result was first obtained by Milne. In problem 4.1 the fact that the $Z^{\frac{7}{3}}$ dependence can be obtained from the Bohr formula (2.7) is stressed. Obviously the coefficient of $Z^{\frac{7}{3}}$ in eqn. (4.20) is a consequence of the self-consistent Thomas–Fermi theory.

We wish to stress also that examination of Fig. 3.1 giving the solutions of the dimensionless Thomas–Fermi equation shows that, in eqn. (4.18), $\phi'(0)$ and x_0 are not functions of Z and N separately, but of (N/Z). Thus, since the Thomas–Fermi theory becomes valid for a non-relativistic Schrödinger equation in which $Z \to \infty$ and $N \to \infty$ such that $(N/Z) \lesssim 1$, we must have in that limit

$$E(Z, N) = Z^{\frac{7}{3}} f\left(\frac{N}{Z}\right) \frac{e^2}{a_0}. \tag{4.21}$$

Using the solutions of the Thomas–Fermi equation in Appendix 3.1, $f(N/Z)$ can be roughly sketched (March and White, 1972) and has the form shown in Fig. 4.1. We shall return to discuss the relevance of this result (4.21) to detailed Hartree and Hartree–Fock studies of atomic binding energies, later in this volume.

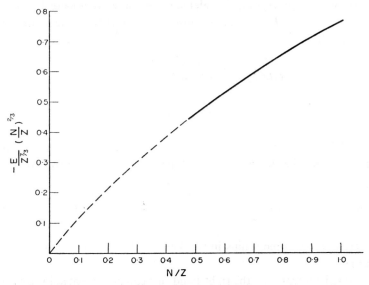

Fig. 4.1. Form of function $f(N/Z)$ in eqn. (4.21). Quantity actually plotted is $-(E/Z^{\frac{7}{3}})(N/Z)^{\frac{2}{3}}$ against (N/Z), with E in units of e^2/a_0 (atomic units).

But we must next derive the basic formula (4.18) for the binding energy.

(a) *Derivation of binding energy formula*

In the Thomas–Fermi theory, we have seen that the three energy terms are explicitly

and

$$
T = c_k \int \{\varrho(\mathbf{r})\}^{\frac{5}{3}} \, d\mathbf{r}, \qquad V_{en} = -Ze^2 \int \frac{\varrho(\mathbf{r})}{r} \, d\mathbf{r} \\[2mm]
V_{ee} = \tfrac{1}{2} \int \varrho(\mathbf{r}) V_e(\mathbf{r}) \, d\mathbf{r}
\tag{4.22}
$$

where $V = -(Ze^2/r) + V_e$ is related to ϱ through the Thomas–Fermi three-halves relation.

One way of proceeding to calculate the energy (compare Hulthén, 1935) is to form dE/dZ at constant N/Z. We start from the mathematical relation

$$
\begin{aligned}
\left(\frac{dE}{dZ}\right)_{N/Z} &= \frac{\partial E}{\partial N} \frac{dN}{dZ} + \frac{\partial E}{\partial Z} \\[2mm]
&= \frac{\partial E}{\partial N} \frac{N}{Z} - e^2 \int \frac{\varrho(\mathbf{r})}{r} \, d\mathbf{r},
\end{aligned}
\tag{4.23}
$$

since N/Z is assumed constant. Then it follows from the result discussed above for the chemical potential that $\partial E/\partial N = -\lambda = V(r_0)$[†] and

$$
\left(\frac{dE}{dZ}\right)_{N/Z} = -\frac{(Z-N)e^2}{r_0} \frac{N}{Z} - e^2 \int \frac{\varrho(\mathbf{r})}{r} \, d\mathbf{r},
\tag{4.24}
$$

where r_0 is the ionic radius in the Thomas–Fermi theory, introduced earlier.

Now the integral on the right-hand side can be evaluated by arguing that the nuclear charge is sitting in the potential created at the nucleus

[†] Using eqn. (4.13) and the fact that $\varrho(r_0) = 0$ at the ionic radius r_0, the electrostatic potential just outside r_0 is $(Z-N)e/r_0$ and is equal to λ/e.

by the electron cloud, or alternatively by using eqns. (3.9) aod (3.10);

$$\int \frac{\varrho(\mathbf{r})}{r} \, d\mathbf{r} = \frac{Z}{b} [\phi'(x_0) - \phi'(0)]. \qquad (4.25)$$

From Fig. 3.1, showing both the positive ion and the neutral atom solutions of the Thomas–Fermi equation, we see immediately that

$$x_0 \phi'(x_0) = -\left(1 - \frac{N}{Z}\right). \qquad (4.26)$$

Thus we can readily rewrite $(dE/dZ)_{N/Z}$ in the form

$$\left(\frac{dE}{dZ}\right)_{N/Z} = 4\left[\frac{2}{9\pi^2}\right]^{\frac{1}{3}} \left[\phi'(0) + \left(1 - \frac{N}{Z}\right)^2 \frac{1}{x_0}\right] Z^{\frac{4}{3}} \frac{e^2}{a_0}. \qquad (4.27)$$

This expression can now be integrated with respect to Z, at constant N/Z, when we obtain the result (4.18).

4.3. Relation between different energy terms for heavy atoms

We have the well known form (4.17) of the virial theorem for a system in equilibrium under the influence solely of Coulomb forces. Also we know

$$V = V_{en} + V_{ee} \qquad (4.28)$$

and as we have already remarked

$$V_{en} = Ze\chi_e(0) \qquad (4.29)$$

where $\chi_e(0)$ is the electrostatic potential at the nucleus due to the electron cloud.

We wish now to relate this to V_{ee}. This we do by using the fact that the kinetic energy has the form (see eqns. (4.6) and (3.4))

$$T = c_k \int \{\varrho(\mathbf{r})\}^{\frac{5}{3}} \, d\mathbf{r} = -\tfrac{3}{5} \int \varrho(V - E_f) \, d\mathbf{r}. \qquad (4.30)$$

Thus the total energy E has the alternative form

$$E = T + V = -\frac{2}{5} Ze^2 \int \frac{\varrho(\mathbf{r})}{r} \, d\mathbf{r} - \frac{1}{10} \int V_e(\mathbf{r}) \varrho(\mathbf{r}) \, d\mathbf{r} \frac{3}{5} NE_f$$

$$(4.31)$$

and because of the virial theorem $E = V/2$ we can then derive the relation

$$\frac{1}{2} \int V_e(\mathbf{r}) \varrho(\mathbf{r}) \, d\mathbf{r} = \frac{1}{7} Ze^2 \int \frac{\varrho(\mathbf{r})}{r} \, d\mathbf{r} + \frac{6}{7} NE_f. \qquad (4.32)$$

But for a neutral atom we have already seen that $E_f = 0$ in the Thomas–Fermi theory. Then we obtain immediately the nice relation between electron–nuclear potential energy V_{en} and electron–electron potential energy V_{ee}

$$V_{ee} = -\frac{1}{7} V_{en} \qquad (4.33)$$

We must stress that such a relation is only precise in the limit $Z \to \infty$ in a non-relativistic framework.

4.4. Foldy's calculation of atomic binding energies

Foldy uses the calculations of Dickinson (1950), discussed briefly below, for the electrostatic potential $\chi_e(0)$ at the nucleus due to the electron cloud. These values were obtained from Hartree field calculations for atoms between He $(Z = 2)$ and Hg $(Z = 80)$. Foldy observed that if one plots the logarithm of the electrostatic potential at the nucleus against ln Z, then a straight line is obtained (see Fig. 4.2), which can be represented by

$$-e\chi_e(0) = \frac{6}{5} Z^{\frac{7}{5}} \frac{e^2}{a_0}. \qquad (4.34)$$

To relate $\chi_e(0)$ to the total binding energy of an atom, use is made of Feynman's theorem (Feynman, 1939). This theorem states that the partial derivative of an energy eigenvalue of a system with respect to a

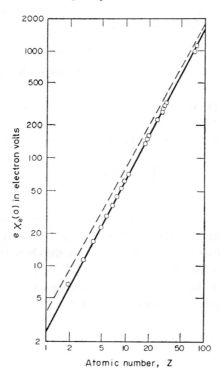

FIG. 4.2. Electrostatic potential $\chi_e(0)$ at nucleus, due to electron cloud, from Hartree self-consistent field calculations, as function of atomic number Z. (Note that this is a log–log plot.) Solid curve—best fit to points, given by eqn. (4.34). Dashed curve is Thomas–Fermi result (see eqn. (4.43)). (After Foldy, 1951.)

parameter occurring in the Hamiltonian of the system is the same as the average value of the partial derivative of the Hamiltonian operator with respect to that parameter.

We now apply this theorem to an ion with atomic number Z and with N electrons. We then find

$$\frac{\partial E}{\partial Z} = e\chi_e(0) \qquad (4.35)$$

where E is the binding energy of the system. In the differentiation, N is held constant.

If, after carrying out the differentiation, N is set equal to Z, then $\chi_e(0)$ can be identified with that calculated by Dickinson for neutral atoms.

So far, no approximations are involved. But now the difference in binding energies between an atom of atomic number $Z-1$ and the singly ionized atom of atomic number Z is approximated by

$$\int_{Z-1}^{Z} e\chi_e(Z) \, dZ. \tag{4.36}$$

Here, of course, we are now going to use the above result for χ_e as an interpolation formula between integral values of Z. The total charge of the electronic cloud is then kept equal to the charge on the nucleus as the atomic number varies from $Z-1$ to Z.

Hence we have that the difference in binding energies between neutral atoms with atomic number $Z-1$ and Z is given approximately by

$$E(Z)-E(Z-1) = \int_{Z-1}^{Z} e\chi_e(Z) \, dZ + I_Z. \tag{4.37}$$

The quantity I_Z in the above equation is the first ionization potential of the atom of atomic number Z.

It is then a straightforward matter, by summation, and using the explicit form for $\chi_e(Z)$ to obtain the result

$$E(Z) = \int_{2}^{Z} e\chi_e(Z) \, dZ + \sum_{z=3}^{Z} I_z + E(2) \tag{4.38}$$

Here, following the discussion of Foldy, the binding energy of He has been used to determine the lower limit of the summation. The first ionization potentials are known for almost all the elements and hence $E(Z)$ can be calculated from the above formula. The results are shown in Table 4.1, along with the Thomas–Fermi values found from eqn. (4.20).

In spite of the apparent dependence as $Z^{\frac{12}{5}}$,[†] there is every reason to believe that the Thomas–Fermi dependence of E on Z, namely $E \propto Z^{\frac{7}{3}}$ is the correct dependence for large Z in a nonrelativistic theory. As dis-

TABLE 4.1 CALCULATED AND EXPERIMENTAL
ATOMIC BINDING ENERGIES (in eV)

Atomic number Z	Hartree results calculated by Foldy	Thomas–Fermi results	Experiment
2	78·63	105·0	78·63
3	202·1	270·1	202·49
4	400·3	529·0	397·15
5	676·8	890·5	667·59
6	1041	1363	1024·87
7	1506	1952	1473·37
8	2068	2666	2032·98
10	3535	4486	
20	$1·833\times10^4$	$2·262\times10^4$	
30	$4·811\times10^4$	$5·824\times10^4$	
40	$9·590\times10^4$	$1·140\times10^5$	
50	$1·637\times10^5$	$1·918\times10^5$	
60	$2·532\times10^5$	$2·934\times10^5$	
70	$3·663\times10^5$	$4·207\times10^5$	
80	$5·049\times10^5$	$5·744\times10^5$	
90	$6·705\times10^5$	$7·561\times10^5$	

cussed by March and White (1972), it appears that one has just not got to large enough Z at Hg ($Z = 80$) to reveal this dependence.

Of course, in practice, relativistic effects come in at large Z, and alter the Z dependence of E. Thus, the results must be corrected for these relativistic effects and this is considered in Chapter 9.

[†] This follows from eqn. (4.38) for large Z, since $\chi_e(Z) Z^{\frac{7}{5}} \propto$ from eqn. (4.34).

(a) *Electrostatic potential at nucleus due to electronic charge cloud*

We have seen that the electron–nuclear potential energy in an atom or ion is given quite generally by

$$V_{en} = \int \varrho(r) \left(\frac{-Ze^2}{r} \right) d\mathbf{r} \tag{4.39}$$

where $\varrho(\mathbf{r})$ is the electron density. If we denote by $\chi_e(r)$ the electrostatic potential due to the charge cloud $-e\varrho(\mathbf{r})$, then evidently at position $\mathbf{r}'$,

$$\chi_e(\mathbf{r}') = \int \frac{(-)e\varrho(\mathbf{r})}{|\mathbf{r}-\mathbf{r}'|} d\mathbf{r}.$$

Since

$$\chi_e(0) = -\int \frac{e\varrho(r)}{r} d\mathbf{r} \tag{4.40}$$

(see also eqn. (4.29)) we can write

$$V_{en} = Ze\chi_e(0), \tag{4.41}$$

which is a quite general result, applicable to the exact electron density $\varrho(\mathbf{r})$.

However, to relate V_{en} to the total energy of an atom or an ion, we must use some sort of approximate theory (e.g. Hartree theory yields (4.38)).

We find, in the Thomas–Fermi approximation, using eqn. (4.33)

$$E = \tfrac{3}{7} V_{en} = \tfrac{3}{7} Ze\chi_e(0). \tag{4.42}$$

From eqn. (4.20), the solution of the dimensionless Thomas–Fermi equation for the isolated atom yields immediately

$$\chi_e(0) = -1{\cdot}79Z^{\frac{4}{3}} \frac{e}{a_0}. \tag{4.43}$$

It is perfectly clear from the statistical assumptions involved that these relations would only become exact in a non-relativistic case as $Z \to \infty$.

However, as remarked above, Dickinson has examined $\chi_e(0)$ for neutral isolated atoms from Hartree theory and he has found the dependence on Z to be better represented by $Z^{\frac{7}{5}}$ through the range $Z = 10$ to 92. Some discussion of the significance of the related results $E \propto Z^{\frac{12}{5}}$ is given by March and White (1972). Foldy's results for $\chi_e(0)$ were shown earlier in Fig. 4.2.

(b) *Effective total charge in Hartree theory*

Let us define the electronic charge enclosed by a sphere of radius r in the atom or ion as

$$Q(r) = - \int_0^r 4\pi e \varrho(r) r^2 \, dr. \qquad (4.44)$$

Clearly, in a neutral atom $Q(r) \to -Ze$ as $r \to \infty$.

The electrostatic potential $\chi_e(r)$ due to this electronic cloud is determined by the Poisson equation

or

$$\left.\begin{array}{c} \dfrac{1}{r^2} \dfrac{d}{dr} \left[r^2 \dfrac{d}{dr} \chi_e(r) \right] = 4\pi e \varrho \\[3mm] -\dfrac{d}{dr} (\chi_e) = \dfrac{Q(r)}{r^2} . \end{array}\right\} \qquad (4.45)$$

Thus we find

$$\chi_e(r) = \int_r^\infty \frac{Q(r)}{r^2} \, dr, \qquad (4.46)$$

which evidently satisfies the boundary condition $\chi_e \to 0$ as $r \to \infty$ like $(-Ze/r)$ for the neutral atom.

The electrostatic potential at the nucleus due to the electronic charge distribution is evidently given by

$$\chi_e(0) = \int_0^\infty \frac{Q(r)}{r^2} \, dr. \qquad (4.47)$$

The Thomas–Fermi approximation specifies $Q(r)$ through eqns. (3.4) and (3.10) but since $\varrho(r)$ is proportional (erroneously) to $r^{-\frac{3}{2}}$ in this model at small r,[†] we have $Q(r) \propto r^{\frac{3}{2}}$ at small r, instead of being equal to $-e\frac{4}{3}\pi r^3 \varrho(0)$. Thus, we overestimate the small r contribution to the above integral and hence the electrostatic potential $\chi_e(0)$ is too negative. This results, through eqn. (4.42) in binding energies which are too large. A correction for this was originally proposed by Scott (1952; see problem 7.4). The energy $\propto Z^{\frac{7}{3}}$ is raised by a term proportional to Z^2. The next contribution in such a series is $Z^{\frac{5}{3}}$, and this term contains the Dirac exchange energy (see Chapter 7) as a major contribution.

Problems

1. Using the Bohr formula (2.7) what is the total energy/closed shell for independent electrons moving in a Coulomb field.

   Suppose we use this oversimplified model for $\mathscr{N}$ closed shells, for a neutral atom. Write down the energy (sum of eigenvalues) for $\mathscr{N}$ closed shells. Develop an explicit relation between $\mathscr{N}$ and Z for Z (and $\mathscr{N}$) very large. Show that the sum of the eigenvalues is proportional to $Z^{\frac{7}{3}}$ for large Z. Compare the coefficient of $Z^{\frac{7}{3}}$ with the self-consistent Thomas–Fermi value (4.20).

2. Calculate for the ground state of the hydrogen atom the function $Q(r)$ giving the amount of electronic charge inside a sphere of radius r (see eqn. (6.12) below).

   How does $Q(r)$ change when we go to the $2s$ excited state?

3. Evaluate the electrostatic potential at the nucleus due to the charge cloud of the electron in the ground state of the hydrogen atom (see p.26).

4. Demonstrate the equivalence of the result (4.40) and the previous equation with eqns. (4.47) and (4.46) respectively.

---

† This error arises because the Thomas–Fermi approximation is valid only when the potential energy $V(r)$ varies by but a small fraction of itself, over a characteristic electron wavelength. The potential V varies too rapidly near to the nucleus for this to be true.

CHAPTER 5

Other atomic properties

WE have discussed the total energies of atoms and ions at some length in the previous chapter. The electrostatic potential at the nucleus due to the electronic charge cloud in the atom was basic to the discussion of binding energies. This same quantity, as we shall show below, also determines the internal diamagnetic field in an atom, when an external magnetic field is applied. We shall follow this discussion with the Langevin–Pauli theory of orbital diamagnetism in atoms. Finally, in this chapter, we shall deal with linear and angular momentum distributions in atoms and with one-electron eigenvalues.

5.1. Internal diamagnetic field in atoms

Dickinson (1950) has discussed the problem of the induced shielding field at the nucleus for an atom or ion in a magnetic field B.

The physics behind this was exposed by Lamb (1941). The starting point is Larmor's theorem, which states that the motion of the atomic electrons in the magnetic field is the same (neglecting terms in B^2), as that before the establishment of the field, except for the superposition of the Larmor precession. The precessing electric charge may be treated as a loop of current which produces a magnetic field at the nucleus opposing the applied field. This shielding field B_i, although numerically small compared with the external field, constitutes an important correction, when, for example, one attempts to measure nuclear magnetic moments by a resonance experiment.[†]

[†] B_i is in fact only part of the chemical shift—there is also a paramagnetic contribution due to the polarization of the atomic shells by B.

Lamb obtained an expression for the shielding field, and in particular he showed that it depended directly on the electrostatic potential $\chi_e(0)$ at the nucleus due to the atomic electrons. Using the result of the Thomas–Fermi theory for $\chi_e(0)$, Lamb showed that the induced shielding field B_i at the nucleus is related to the external field by

$$(B_i/B) = -0{\cdot}319 \times 10^{-4} Z^{\frac{4}{3}}. \tag{5.1}$$

Lamb checked the above formula for the cases $Z = 19, 20, 26, 29, 37, 55, 74,$ and 80 where $\chi_e(0)$ was explicitly available from Hartree field calculations, and this work was subsequently extended by Dickinson as we discussed in the previous Chapter.

In particular, Dickinson obtained $\chi_e(0)$ for all atoms and singly charged ions for which Hartree or Hartree–Fock[†] calculations were available at that time.

(a) *Derivation of Lamb's result*

We consider an atom with a spherical charge distribution in an external field B. As an element of volume to consider, we choose a ring with axis passing through the nucleus and parallel to B, with cross-section $r d\theta \, dr$ and perimeter $2\pi r \sin \theta$. Thus, the volume of the ring is $2\pi r^2 \sin \theta d \theta \, dr$ and the charge it contains is clearly $\sin\theta \, d\theta \, dr \, Q'(r)/2$.

The rotation of this ring with the Larmor frequency $\omega = eB/2mc$ (cf. eqn. (5.17) below) results in a current

$$di = [\sin \theta \, d\theta \, dr \, Q'(r)/2] eB/4\pi mc. \tag{5.2}$$

It then follows from the Biot–Savart Law that the field dB_i at the nucleus due to this current loop is

$$dB_i = \frac{2\pi \, di \, \sin^2 \theta}{rc}. \tag{5.3}$$

[†] The Hartree–Fock equations are given in eqn. (7.36) below.

If we use the above result for di, we find, with the help of eqn. (4.47),

$$\frac{B_i}{B} = \frac{e}{4mc^2} \int_0^\pi \sin^3 \theta \, d\theta \int_0^\infty \frac{Q'(r)}{r} \, dr$$

$$= \frac{e}{3mc^2} \chi_e(0) \qquad (5.4)$$

which is simply Lamb's result.

It should be remarked that, in a more general case, instead of a scalar coupling to B, there would be a tensor coupling. This would be relevant, for example, in a molecular beam experiment.

5.2. Orbital diamagnetic susceptibility

Before giving a proper quantum-mechanical derivation of the Langevin–Pauli formula for atomic diamagnetism, we will give a classical 'hand-waving' argument which will also serve to expand some of the material presented in the previous section, 5.1.

Again consider the motion of an electron in a circular orbit; this electron behaves like a tiny current loop and has an orbital magnetic dipole moment. If, to be definite, we take the hydrogen atom, we can write down the Newton equation of motion as

$$\text{Force} = \frac{e^2}{r^2} = \frac{mv^2}{r} \qquad (5.5)$$

v^2/r being the acceleration due to motion in a circle of radius r with tangential velocity v. Thus we can write for the angular velocity

$$\omega = \frac{v}{r} = \left\{ \frac{e^2}{mr^3} \right\}^{\frac{1}{2}}. \qquad (5.6)$$

Now the current i, or the rate at which charge passes any given point, may be expressed as

$$i = \frac{e\omega}{2\pi} = \frac{e^2}{2\pi} \left\{ \frac{1}{mr^3} \right\}^{\frac{1}{2}}. \qquad (5.7)$$

Again using the formula for the magnetic effect of a current, we have for the orbital dipole moment μ

$$\mu = iA$$

$$= \pi r^2 \frac{e^2}{2\pi} \left\{ \frac{1}{mr^3} \right\}^{\frac{1}{2}}$$

$$= \frac{e^2}{2} \left\{ \frac{r}{m} \right\}^{\frac{1}{2}}, \tag{5.8}$$

where A is the area of the orbit, namely πr^2.

The orbital angular momentum L is evidently given by (1.3) as

$$L = mvr = mr \left\{ \frac{e^2}{mr} \right\}^{\frac{1}{2}} \tag{5.9}$$

and hence, in these units,

$$\mu = \frac{1}{2} \frac{e}{m} L.\dagger \tag{5.10}$$

Now we examine the effect of applying a magnetic field normal to such a conducting ring. Applying Faraday's Law, we can write

$$\int \mathbf{E}.d\mathbf{l} = -\frac{1}{c} \frac{d}{dt} \int \mathbf{B}.d\mathbf{S} \tag{5.11}$$

which in free space, for a ring of radius $R\ddagger$ becomes

$$E2\pi R = -\frac{\pi R^2}{c} \frac{dB}{dt}. \tag{5.12}$$

A charge of magnitude e in such a conducting ring will be set in motion with a force

$$Ee = -\frac{eR}{2c} \frac{dB}{dt} = \frac{d}{dt}(mv) \tag{5.13}$$

---

† More usually written with L multiplied by $e/2mc$ (see Pauling and Wilson, 1935).
‡ Larmor's theorem is again explicit here.

and hence

$$m\frac{dv}{dt} = -\frac{eR}{2c}\frac{dB}{dt}.$$ (5.14)

Integrating with respect to t, and using the initial condition B, $v = 0$ at $t = 0$ we find

$$mv = -\frac{Re}{2c}B$$ (5.15)

so that application of the magnetic field gives the electron a velocity

$$v = -\frac{1}{2}\frac{Re}{mc}B.$$ (5.16)

Thus, the corresponding angular velocity is given by

$$\omega = -\frac{1}{2}\frac{e}{mc}\mathbf{B},$$ (5.17)

which is the Larmor frequency utilized previously. The angular momentum L is simply

$$L = m\omega R^2$$ (5.18)

or

$$L = -\frac{mR^2}{2}\cdot\frac{e}{m}\frac{B}{c}.$$ (5.19)

But from the relation (5.10) between orbital angular momentum and orbital dipole moment we have for the induced magnetic moment

$$\mu_{induced} = -\frac{1}{4}\left(\frac{e^2}{m}\right)\frac{R^2}{c^2}B.$$ (5.20)

The magnetic moment is seen to be always opposed to the direction of the field and is independent of the sign of the charges involved.

Hence the orbital susceptibility $\chi_{orb} = \mu_{induced}/B$ is given by

$$\chi_{orb} = -\frac{1}{4}\frac{e^2}{mc^2}R^2.$$ (5.21)

Below, we give a correct quantum-mechanical theory of the orbital susceptibility. The above formula is regained, provided that R^2 is replaced by the quantum-mechanical average of r^2. There is also a change in the numerical factor (see eqn. (5.30) below), from $\frac{1}{4}$ to $\frac{1}{6}$. This can be obtained classically by averaging over all directions of B with respect to the axis of the orbit.

(a) *Quantum-mechanical calculation*

To discuss the quantum-mechanical calculation of the orbital susceptibility, we take the Hamiltonian

$$H = \Sigma \frac{\left[\mathbf{p} + \dfrac{e\mathbf{A}}{c}\right]^2}{2m} + V \tag{5.22}$$

as starting point, the vector potential $\mathbf{A}$ representing the applied magnetic field (see Appendix 5.1). In a many-electron system, with an appropriate choice of gauge,[†] we can write

$$\begin{aligned}
H = \sum_{i=1}^{N} \Bigg\{ &-\frac{\hbar^2}{2m}\left(\frac{\partial^2}{\partial x_i^2} + \frac{\partial^2}{\partial y_i^2} + \frac{\partial^2}{\partial z_i^2}\right) \\
&+ \frac{\mathcal{H}e\hbar}{2imc}\left(x_i\frac{\partial}{\partial y_i} - y_i\frac{\partial}{\partial x_i}\right) \\
&+ \frac{\mathcal{H}^2 e^2}{8mc^2}(x_i^2 + y_i^2) + V.
\end{aligned} \tag{5.23}$$

We now use the result that the magnetization is related to the free energy through

$$M = -\frac{\partial F}{\partial \mathcal{H}} \tag{5.24}$$

and, at absolute zero, this reduces to

$$M = -\frac{\partial E}{\partial \mathcal{H}}. \tag{5.25}$$

[†] Actually, with $\mathbf{A} = \frac{1}{2}\mathcal{H} \times \mathbf{r}$, and $\mathcal{H} = (0, 0, \mathcal{H})$, $\mathcal{H}$ denoting the applied field in this section,

Then, we can again use the Feynman theorem that the derivative of the energy with respect to a parameter λ in the Hamiltonian can be obtained from (cf. p. 50)

$$\left\langle \frac{\partial H}{\partial \lambda} \right\rangle = \frac{\partial E}{\partial \lambda}. \tag{5.26}$$

We then find, with $\lambda \equiv \mathcal{H}$, from the above Hamiltonian for the ith term

$$\left\langle \frac{\partial H}{\partial \mathcal{H}} \right\rangle = -\frac{e\hbar}{2imc} \left\langle x_i \frac{\partial}{\partial y_i} - y_i \frac{\partial}{\partial x_i} \right\rangle$$
$$+ \frac{e^2 \mathcal{H}}{4mc^2} \langle x_i^2 + y_i^2 \rangle. \tag{5.27}$$

It is evident that, to define a susceptibility, the left-hand side must be proportional to $\mathcal{H}$. This is true of the second term. We can show, indeed, that the first term on the right-hand side vanishes and using the result that

$$\langle r^2 \rangle = \langle x^2 + y^2 + z^2 \rangle$$
$$= \tfrac{3}{2} \langle x^2 + y^2 \rangle \tag{5.28}$$

we obtain

$$\left\langle \frac{\partial H}{\partial \mathcal{H}} \right\rangle = \frac{e^2 \mathcal{H}}{6mc^2} \langle r^2 \rangle. \tag{5.29}$$

TABLE 5.1 CALCULATED
AND OBSERVED
DIAMAGNETIC SUSCEPTIBILI-
TIES FOR ATOMS AND IONS
$(-\chi_{orb} \times 10^6)$

	Hartree field	Observed
He	1·9	1·9
Na$^+$	5·6	6
K$^+$	17·3	15
Rb$^+$	29·5	22

Hence the orbital susceptibility from (5.25), (5.26) and (5.29), is

$$\chi_{orb} = -\frac{e^2}{6mc^2} \sum_i \langle r_i^2 \rangle. \tag{5.30}$$

Some numerical results for χ using self-consistent field wave functions for atoms and ions are recorded in Table 5.1. There is quite reasonable agreement between theory and experiment.

(b) *Calculation of $\langle r^2 \rangle$ for He atom*

We saw variationally that an approximate wave function for a two-electron ion with nuclear charge Z is given for the ground state by

$$\Psi(\mathbf{r_1 r_2}) = \exp\left(-\left\{Z-\frac{5}{16}\right\}\frac{r_1}{a_0}\right) \exp\left(-\left\{Z-\frac{5}{16}\right\}\frac{r_2}{a_0}\right). \tag{5.31}$$

Shull and Löwdin (1956) pointed out that if we allowed different spins to correspond to a different space orbital then we could write a symmetrized wave function as

$$\Psi(\mathbf{r_1 r_2}) = N[\exp(-ar_1)\exp(-br_2) + \exp(-br_1)\exp(-ar_2)]. \tag{5.32}$$

They obtained a and b by minimizing the energy, and their conclusion was that the minimum energy for the He atom, with $Z = 2$, did not correspond to the choice $a = b = 2 - \frac{5}{16}$, but to unequal values $a = 2 \cdot 183$ and $b = 1 \cdot 189$.

It is a straightforward matter to determine:

   (i) The normalization factor N in the above wave function
   (ii) The mean value of r^2.

The result for N is readily shown to take the form

$$N = \frac{1}{8\sqrt{(2)}\pi\left[\frac{1}{64}(ab)^{-3} + (a+b)^{-6}\right]^{\frac{1}{2}}} \tag{5.33}$$

and inserting the above numerical values we find

$$N^2 = 0 \cdot 503. \tag{5.34}$$

Similarly, it is left as an exercise for the reader to show that

$$\langle r^2 \rangle = 3\pi^2 N^2 \left[\frac{1}{b^3 a^5} + \frac{1}{a^3 b^5} + \frac{512}{(a+b)^8} \right]. \tag{5.35}$$

We then find

$$\langle r^2 \rangle = 1 \cdot 238 a_0^2, \tag{5.36}$$

whereas with $a = b$ we obtain $\langle r^2 \rangle = a_0^2$.

This difference shows that electron–electron correlations, which are responsible for the parameters a and b being different, are actually expanding the charge cloud a little. But this anticipates a more extensive discussion of electron correlation in Chapter 8.

5.3. Linear momentum distribution in atoms

The standard way of obtaining momentum distributions of electrons in atoms is to calculate the momentum wave functions by:

(i) Fourier transform of the coordinate space wave functions. For example, this transform could be obtained numerically from Hartree self-consistent field wave functions.

(ii) Solution of the appropriate wave equation in momentum space.

These two approaches are set out in Appendix 5.2, in general terms.

For heavy atoms, for which such direct calculations of momentum wave functions become laborious, we can use as a cruder alternative the Thomas–Fermi method. We have the relation (3.3) between electron density $\varrho(r)$ and maximum momentum $p_f(r)$. Furthermore, around the point $\mathbf{r}$, we can evidently write that the probability of an electron at $\mathbf{r}$ having momentum between p and $p + dp$ is given by $4\pi p^2 \, dp / \frac{4}{3}\pi p_f^3(\mathbf{r})$.

Hence if we introduce a local momentum distribution function $I_r(p) \, dp$, then we have

$$\left. \begin{aligned} I_r(p) \, dp &= \frac{3p^2 \, dp}{p_f^3(\mathbf{r})} \quad &\text{for} \quad p \leqslant p_f(r) \\ &= 0 \quad &\text{for} \quad p > p_f(r) \end{aligned} \right\}. \tag{5.37}$$

Thus we can write, with Z the total number of electrons equal to $\int_0^\infty \varrho(r)\,4\pi r^2\,dr$, that the probability of an electron having momentum p is

$$I(p)\,dp = \int_{I_r(p)\neq 0} \frac{I_r(p)\,dp\,\varrho(r)4\pi r^2\,dr}{Z}. \tag{5.38}$$

For a given r, as can be seen from Fig. 5.1(a), p goes from 0 to $p_f(r)$. For a given momentum p, r goes from 0 to $r(p)$, where $r(p)$ is defined by the equation of boundary curve. Thus we have, using eqn. (3.3),

$$I(p)\,dp = \frac{32\pi^2}{3h^3Z}\,r^3(p)p^2\,dp. \tag{5.39}$$

The form of the results is shown schematically in Fig. 5.1(b). The model has the merit that $I(p)$ can be displayed in a dimensionless form applicable to all atoms (Coulson and March, 1950; Konya, 1951).

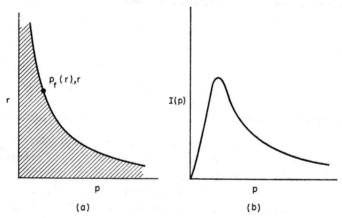

FIG. 5.1. (a) The (r, p) plane for the Thomas–Fermi atom. (b) Momentum distribution function $I(p)$ for Thomas–Fermi atom (schematic).

(a) *Mean linear momentum*

It is now a straightforward matter to calculate the mean momentum of an electron both in the Thomas–Fermi theory and from self-consistent field calculations. The latter will, of course, reflect the Periodic Table in detail, whereas the Thomas–Fermi theory will yield results which vary smoothly with atomic number.

Figure 5.2 shows the mean momentum $\bar{p}$, as a function of atomic number Z. Explicitly $\bar{p}$ is given by

$$\bar{p} = \int_0^\infty pI(p)\,dp \tag{5.40}$$

and in the Thomas–Fermi theory this varies as $Z^{\frac{2}{3}}$ $\Big($the kinetic energy/

electron varying as $Z^{\frac{4}{3}}\Big)$. The smooth curve is obtained by calculating the momentum distribution function $I(p)$ from the solution of the dimensionless Thomas–Fermi equation representing the neutral atom.

The other results shown are obtained from approximate wave functions first discussed by Morse, Young and Haurwitz (1935; see Appendix

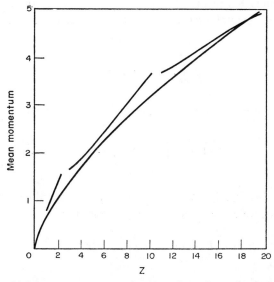

FIG. 5.2. Mean momentum $\bar{p}$ as function of atomic number Z. Smooth curve is result of Thomas–Fermi model. Other results obtained from wave functions of Morse, Young and Haurwitz (see Appendix 6.1).

6.1 for their explicit forms). Naturally, the wave mechanical results reflect periodic structure, as already remarked.

The agreement here is reasonably satisfactory and, as Fig. 5.2 indicates, the accuracy of the Thomas–Fermi result improves as Z becomes larger.

5.4. Orbital angular momentum

Fermi, in his original paper which is included in this volume (on p. 207) discussed the angular momentum distribution in atoms.

We shall briefly record here the work of Jensen and Luttinger (1952). These workers, instead of trying to estimate the 'first appearance' of an electron with a given orbital angular momentum, which, while interesting, is somewhat arbitrary, focused on a well-defined quantity, namely the square of the orbital angular momentum $\langle L^2 \rangle$.

The Thomas–Fermi method allows us to calculate the number of electrons with orbital angular momentum between L and $L+dL$, say $I(L)\,dL$, such that

$$\int_0^\infty I(L)\,dL = Z \tag{5.41}$$

for a neutral atom. Then the mean value of L^2 is evidently

$$\langle L^2 \rangle = \int L^2 I(L)\,dL/Z. \tag{5.42}$$

On the other hand, from our knowledge of the filling of atomic energy levels, we can evidently write (see eqn. (1.41))

$$\hbar^{-2} \langle L^2 \rangle = \sum_i l_i(l_i+1)/Z, \tag{5.43}$$

the sum being taken over the Z electrons.

To set up the orbital angular momentum distribution function $I(L)$, we note that the number of electrons with momentum **p** at position **r**, in the element of phase space of volume $d\mathbf{p}\,d\mathbf{r}$ is given by (see section 3.2)

$$d\varrho(\mathbf{pr}) = \frac{2}{h^3} d\mathbf{r}\,d\mathbf{p} \quad \text{if} \quad p \leqslant p_f(r) \Bigg\} .$$

$$\text{if} \quad p > p_f(r) \Bigg\} \tag{5.44}$$

It is obvious that if we integrate this result over p we find the basic Thomas–Fermi relation (3.3) between density and maximum momentum p_f.

Let us now consider the number of electrons with angular momentum greater than L, namely $I_>(L)$ such that

$$I_>(L) = \int_L^\infty I(L)\,dL. \tag{5.45}$$

Now quite generally the orbital angular momentum is defined by eqn. (1.3) and therefore if θ is the angle between the vectors $\mathbf{r}$ and $\mathbf{p}$ we can write

$$L^2 = r^2 p^2 \sin^2 \theta. \tag{5.46}$$

Using the result above for $d\varrho(\mathbf{pr})$, we can evidently write that

$$I_>(L) = \int d\varrho(\mathbf{pr}) \tag{5.47}$$

where the limits of the r and p integrations are given by the inequalities

$$rp \sin \theta \geqslant L; \quad p \leqslant p_f(r). \tag{5.48}$$

Integrating over angles, and over p, we then find

$$\hbar^3 I_>(L) = \frac{4}{3\pi} \int [r^2 p_f^2(r) - L^2]^{\frac{3}{2}} \frac{dr}{r} . \tag{5.49}$$

It is clear that the integral over r is to be taken over the region in which the square root is real. Thus, by differentiating with respect to L we obtain the distribution function $I(L)$ as

$$I(L) = -\frac{dI_>(L)}{dL} = \frac{4L}{\pi \hbar^3} \int [r^2 p_f^2(r) - L^2]^{\frac{1}{2}} \frac{dr}{r} \tag{5.50}$$

which is essentially the result given by Fermi (1928; see reprint 3 in this volume).

We can immediately form $\langle L^2 \rangle$ and we find

$$\hbar^3 \langle L^2 \rangle = \frac{8}{15\pi Z} \int_0^\infty \frac{dr}{r} [r p_f(r)]^5. \tag{5.51}$$

In terms of the solution ϕ of the dimensionless Thomas–Fermi equation, appropriate to neutral atoms, we find

$$\langle L^2 \rangle = \hbar^2 Z^{\frac{2}{3}} \frac{2}{5} \left(\frac{3\pi}{4}\right)^{\frac{2}{3}} \int_0^\infty \frac{dx}{x} (x\phi)^{\frac{5}{2}}. \tag{5.52}$$

The integral has the value 0.370 as can be verified by numerical integration, and hence

$$\langle L^2 \rangle = 0.262 Z^{\frac{2}{3}} \hbar^2. \tag{5.53}$$

This quantity is plotted in Fig. 5.3, where the empirical value of $\langle L^2 \rangle$ as calculated from the known ground state configurations of the elements is also shown.

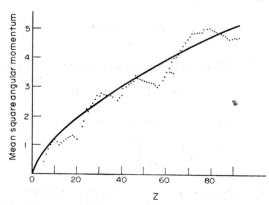

FIG. 5.3. $\langle L^2 \rangle$ versus Z. Smooth curve is result of Thomas–Fermi model, given in eqn. (5.53). Other results are as calculated from the known ground state configurations of the elements (after Jensen and Luttinger, 1952).

As expected, the accuracy with which the Thomas–Fermi theory represents an average of the properties of atoms again improves with increasing Z. It is clear that we could not expect more than such an average representation, for all effects due to closing of shells are averaged out in the Thomas–Fermi theory.

5.5. Eigenvalues for potentials related to Thomas–Fermi self-consistent fields

One of the difficulties underlying the statistical theory is that it yields a 'density of states curve' for bound state energies less than zero, whereas of course the energy levels are discrete (problem 3.3).

Nevertheless, the Thomas–Fermi self-consistent fields have been used by a number of workers to determine atomic term values. A rather complete investigation has been made by Latter (1955), and we shall record here his main results.

Latter's work uses both Thomas–Fermi and Thomas–Fermi–Dirac[†] results, but he modifies the potential in such a way that at large distances the field becomes that of a unit positive charge. Some account is thereby taken of the fact that, in the Thomas–Fermi theory, the field created by the charge cloud of the electron under discussion is included in the self-consistent field. The potential energy used for the Thomas–Fermi case was therefore

$$V(r) = -\frac{Ze^2}{r}\phi(x) \quad \text{if} \quad V(r) < -\frac{e^2}{r} \left.\begin{array}{c}\\\\\end{array}\right\}$$
$$= -\frac{e^2}{r} \quad \text{otherwise.} \tag{5.54}$$

Thus, no modification of the field for self-interaction has been made in the interior of the atom, but over most of this region it seems that this should be relatively unimportant. Latter approximated the function $\phi(x)$ which satisfies the dimensionless Thomas–Fermi equation

[†] The Thomas–Fermi–Dirac relation is given in eqn. (7.22).

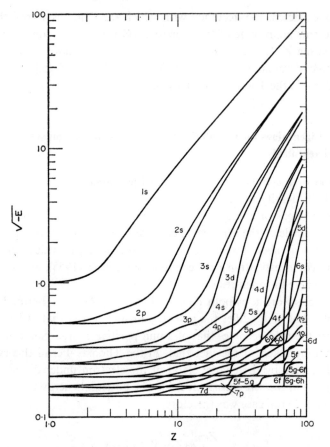

FIG. 5.4. Atomic term values from modified Thomas–Fermi potential.
Quantity actually plotted is square root of term values versus atomic
number Z (after Latter, 1955). Energies in Rydbergs.

by the analytic form

$$\phi(x) = \left[1 + 0 \cdot 02747x^{\frac{1}{2}} + 1 \cdot 243x - 0 \cdot 1486x^{\frac{3}{2}} \right.$$
$$\left. + 0 \cdot 2302x^2 + 0 \cdot 007298x^{\frac{5}{2}} + 0 \cdot 006944x^3 \right]^{-1} \quad (5.55)$$

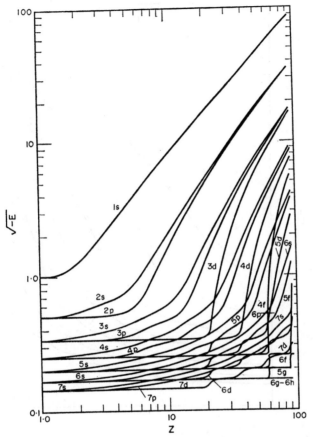

FIG. 5.5. Same as Fig. 5.4, except that modified Thomas–Fermi–Dirac potential (i.e. including exchange) is used (after Latter, 1955).

which yields for small x at least the correct form of series expansion given by Baker (see problem 3.1) while for large x it yields the asymptotic form (3.14).

Latter has constructed an extensive table of term values from $1s$ to $7d$, for a range of values of atomic number Z sufficient to permit easy interpolation. For the precise numerical data, reference may be made to this table, but an indication of the main features of Latter's find-

6*

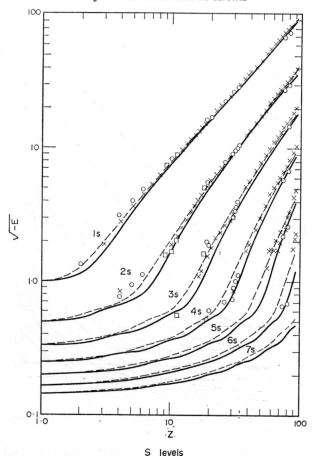

S levels

FIG. 5.6. Comparison of results of Fig. 5.4 (solid line) and 5.5 (dashed line) with Hartree (circles) and Hartree–Fock (squares) calculations, and with experimental results (crosses) Only *s* states shown (after Latter, 1955).

ings is given by Fig. 5.4, where the square roots of the term values are plotted against Z.

Figure 5.5 shows similar results including exchange by modifying the Thomas–Fermi–Dirac potentials.

Finally Figs. 5.6 and 5.7 show a comparison of these results with

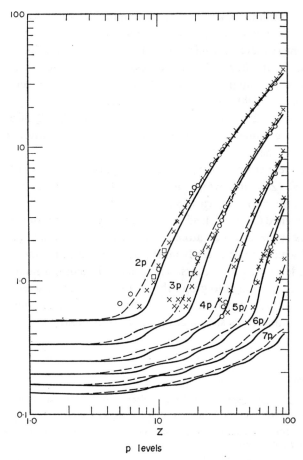

Fig. 5.7. Same as Fig. 5.6, but for *p* states (after Latter, 1955).

Hartree and Hartree–Fock calculations and experimental ionization energies, these two figures showing *s* and *p* levels respectively. Similar results for *d* and *f* levels are given by Latter. For large atomic numbers, it is obvious that relativistic effects become important and must be incorporated before it is reasonable to expect accurate agreement with experiment (see Chapter 9).

We only wish to make one further comment on these results, which concerns the somewhat elementary point as to the order in which one-electron energy levels are filled in atoms. In a Coulomb field, we would, of course, fill the K, L, M shells successively. The closed shell of principal quantum number n holds $2n^2$ electrons and hence shells close, for a Coulomb field, for 2, 10, 28, etc., electrons.

In general, the most realistic definition of the closing of a shell is the complete filling up of a group of energy levels relatively isolated from other energy levels.

Now in real atoms, the closed shell configurations occur of He ($Z = 2$) and Ne ($Z = 10$), which fit with the Coulomb field scheme. But the next rare gas is Ar, with 18 electrons. The reason is clear; namely that a screened Coulomb field such as given by a Hartree or a Thomas–Fermi theory leads for an atomic number $Z = 18$ to the filling of a $3p$ subshell which is rather far from the $3d$ level. Also, for larger atomic number, level crossing occurs. The principal quantum number is no longer precise enough to group the energy levels; for a screened Coulomb field these depend on n and the orbital angular momentum quantum number, as shown clearly in Figs. 5.4 and 5.5.

Problems

1. Calculate the average value of p^2 for an electron in the Thomas–Fermi atom, by using the momentum distribution function $I(p)$ of section 5.3.

2. Make a rough sketch of the Thomas–Fermi angular momentum distribution function $I(L)$ of section 5.4.

3. Calculate the shape of the Compton profile $J(q)$ defined in eqn. (6.22) from the Thomas–Fermi atom result for $I(p)$ given in section 5.3. Show that $J(q)$ in this approximation exhibits a cusp at $q = 0$, in contrast to the correct wave-mechanical forms (cf. Fig. 6.6b).

4. Show that for a harmonic oscillator, the integral equation (A5.2.6) for the momentum wave function reduces to a differential equation. Hence obtain the wave function explicitly in momentum space for the ground state of the oscillator.

5. For a Morse potential, show that the momentum wave function satisfying the integral equation (A5.2.6) can be derived more readily from a difference equation.

CHAPTER 6

X-ray scattering and electron densities in atoms

A CHECK of wave-mechanical self-consistent field calculations on both atoms and molecules can be made by using experimental results for X-ray scattering from gases.

One advantage of gases is that, unlike crystals, the thermal vibrations of the atoms are relatively unimportant and it turns out in most cases that one can make a significant comparison with wave mechanical calculations for non-vibrating atoms.

If we include the molecular case (only the rare gas atoms can otherwise be used) then it should be stressed that by dealing with the scattering from an assembly of randomly oriented molecules, one inevitably loses information and it might be questioned whether any useful results can be obtained.

But Debye showed at an early stage (1915) that this was indeed so. To see this, let us discuss the coherent scattering of X-rays, in terms of the electronic density distribution $\varrho(\mathbf{r})$ which we wish to test. For molecules though, it will not prove possible to extract $\varrho(\mathbf{r})$ from gas scattering experiments, but rather we must calculate $\varrho(\mathbf{r})$ from self-consistent field procedures and then test the theory against experiment.

In comparing with experiment, it is necessary to consider also the incoherent scattering, in which there is a change in wavelength. This, in contrast with the coherent scattering, is a very complicated functional of the electron density.[†] Fortunately it turns out that, while the

---

[†] As shown below, the incoherent scattering is much more intimately related to the momentum density than to $\varrho(\mathbf{r})$.

incoherent scattering is harder to calculate accurately, in many instances it is not difficult to get results to useful accuracy, the incoherent scattering often being less important than the coherent scattering.

Wave-mechanically we ought to formulate the entire scattering problem, both the coherent and the incoherent. However, since our prime interest is in gaining information on the electron density distribution, we shall content ourselves with a brief sketch of the theory of scattering by a charge cloud of density $\varrho(\mathbf{r})$. The argument given below is far from rigorous, but is designed to make clear the general features of the coherent scattering formula.

6.1. Coherent scattering

First of all, we will calculate the angular distribution of X-ray intensity scattered by an actual molecule, free and isolated in space.

Figure 6.1 shows the situation under discussion. An X-ray beam is incident in the direction defined by the unit vector $\mathbf{s_0}$. We shall consider radiation scattered in the direction defined by the unit vector $\mathbf{s}$.

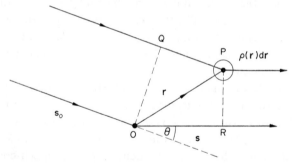

FIG. 6.1. Shows X-ray scattering, as discussed in section 6.1.

The path difference between the rays shown is $OR-QR$

$$= \mathbf{r}.\mathbf{s} - \mathbf{r}.\mathbf{s_0}. \tag{6.1}$$

Since we have for the phase difference

$$\text{Phase difference} = \frac{2\pi}{\lambda} \times \text{path difference} \tag{6.2}$$

where λ is the wavelength of the incident radiation, it follows that

$$\text{Phase difference} = \frac{2\pi}{\lambda} \mathbf{r} . (\mathbf{s} - \mathbf{s}_0) \qquad (6.3)$$

Hence the scattered amplitude from an element $d\mathbf{r}$ of electron density $\varrho(\mathbf{r})$ is given by

$$\text{Scattered amplitude} \propto \exp\left\{\frac{2\pi i}{\lambda} \mathbf{r} . (\mathbf{s} - \mathbf{s}_0)\right\} \varrho(\mathbf{r}) \, d\mathbf{r}$$

$$= \exp\{ik\mathbf{S} . \mathbf{r}\} \varrho(\mathbf{r}) \, d\mathbf{r} \qquad (6.4)$$

where $k = 2\pi/\lambda$ and $\mathbf{S} = \mathbf{s} - \mathbf{s}_0$. Thus, the total scattered amplitude is obtained by integration over the whole electron density distribution as

$$\text{Total scattered amplitude} \propto \int \varrho(\mathbf{r}) \exp(ik\mathbf{S} . \mathbf{r}) \, d\mathbf{r}. \qquad (6.5)$$

Let us write the atomic scattering factor f from an electron distribution of density $\varrho(\mathbf{r})$ as ($\mathbf{K} = k\mathbf{S}$)

$$f(\mathbf{K}) = \int \varrho(\mathbf{r}) \exp(i\mathbf{K} . \mathbf{r}) \, d\mathbf{r} \qquad (6.6)$$

and in the case of a closed shell atom like Ne, $\varrho(r)$ is spherically symmetrical (see Unsöld's theorem in, for example, Pauling and Wilson, 1935). The angular integrations can then be performed by taking the direction of $\mathbf{K}$ as the polar axis, and using spherical polar coordinates r, θ, ϕ. Then the volume element becomes $r^2 \sin\theta \, dr \, d\theta \, d\phi$, and since $\mathbf{K} . \mathbf{r} = Kr \cos\theta$ we can integrate over ϕ from 0 to 2π to obtain

$$f(K) = 2\pi \int_0^\infty dr \, r^2 \varrho(r) \int_0^\pi d\theta \exp(iKr \cos\theta) \sin\theta. \qquad (6.7)$$

Putting $x = \cos\theta$, the integration over θ can be carried out, with the result

$$f(K) = \int_0^\infty \frac{\sin Kr}{Kr} \varrho(r) 4\pi r^2 \, dr \qquad (6.8)$$

which could have been obtained directly by using Bauer's expansion for a plane wave in a series of spherical waves, namely

$$\exp(i\mathbf{K}\cdot\mathbf{r}) = \sum_{l=0}^\infty (2l+1) \, i^l j_l(Kr) \, P_l(\cos\theta) \qquad (6.9)$$

where j_l represents the spherical Bessel function of order l while $P_l(\cos\theta)$ is the lth Legendre polynomial. The spherical symmetry of the electron density $\varrho(r)$ removes all the terms by orthogonality of the Legendre polynomials except the term $l = 0$. Using the explicit form of the zeroth order spherical Bessel function, namely

$$j_0(Kr) = \sin Kr / Kr \qquad (6.10)$$

we regain the result (6.8).

(a) *Scattering factor for ground state of atomic hydrogen*

As the simplest example of the scattering formula (6.6), let us work out the atomic scattering factor for hydrogen in its ground state.

In practice one measures the scattering from molecular hydrogen and we give a plot of $f^2(K)$ for this case in Fig. A6.2.1, p. 155.

The normalized ground-state wave function of atomic hydrogen is readily shown from eqn. (2.1) to be

$$\psi(r) = \frac{1}{(\pi a_0^3)^{\frac{1}{2}}} \exp(-r/a_0) \qquad (6.11)$$

and hence the electron density $\varrho(r)$ has the form

$$\varrho(r) = \frac{1}{\pi a_0^3} \exp(-2r/a_0). \qquad (6.12)$$

We require then

$$f(K) = \int_0^\infty \frac{1}{\pi a_0^3} e^{-2r/a_0} \frac{\sin Kr}{Kr} 4\pi r^2 \, dr$$

$$= \text{Imaginary part} \left[\frac{4}{Ka_0^3} \int_0^\infty e^{-2r/a_0} e^{iKr} r \, dr \right]$$

$$= \text{Imaginary part} \left[\frac{4}{Ka_0^3} \int_0^\infty e^{-r[(2/a_0)-iK]} r \, dr \right] \quad (6.13)$$

Using the result

$$\int_0^\infty x e^{-\alpha x} \, dx = \alpha^{-2} \quad (6.14)$$

we find

$$f(K) = \text{Im} \frac{4}{a_0^3 K} \left[\frac{1}{\left(\dfrac{2}{a_0} - iK \right)^2} \right]$$

$$= \frac{16}{(4 + K^2 a_0^2)^2}. \quad (6.15)$$

In this formula, we remind the reader that $K = 4\pi \sin(\theta/2)\lambda$ where θ is the scattering angle (see Fig. 6.1) and λ is the X-ray wavelength.

The result (6.15) is plotted schematically in Fig. 6.2 and we wish to stress two points here.

(i) $f(0) = 1$.

(ii) $f(K)$ at large $K \sim \dfrac{16}{K^4 a_0^4}$.

The first point is a special case of the general result, obtained by putting $K = 0$ in eqn. (6.6), namely

$$f(0) = \int \varrho(\mathbf{r}) \, d\mathbf{r} = Z \quad (6.16)$$

where $\varrho(\mathbf{r})$ integrates, of course, to the total number of electrons in the atomic charge cloud. In the case of a neutral atom, this is simply the atomic number Z.

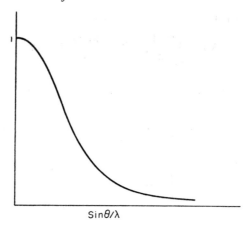

Sinθ/λ

FIG. 6.2. X-ray scattering factor for atomic hydrogen as given by eqn. (6.15) (schematic only). Here 2θ is the scattering angle.

The second point, the slow decay of the atomic scattering factor, follows from the fact that the slope of $\varrho(r)$ at the origin is finite. When this is so, it follows from the theory of Fourier transforms (see, for example, Lighthill, 1958) that the Fourier transform always falls off like K^{-4} (Goscinski and Linder, 1970), and the coefficient depends on the slope $(\partial\varrho/\partial r)_{r=0}$. This is interesting, because, as we show below in Chapter 8, we can get a precise relation between $(\partial\varrho/\partial r)_{r=0}$ and the electron density $\varrho(0)$ at the nucleus.

(b) *Scattering from approximate self-consistent field electron distributions*

McWeeny (1951) has made extensive calculations of atomic scattering factors using analytical wave functions of Morse, Young, Haurwitz form (see Appendix 6.1).

McWeeny considers separately scattering factors for $1s$, $2s$, and from non-spherical $2p$ orbitals, in each case the scattering factor being calculated from the density $\varrho_i(\mathbf{r})$ via

$$f_i = \int \varrho_l(\mathbf{r}) \exp(ik\mathbf{S} \cdot \mathbf{r}) \, d\mathbf{r}. \qquad (6.17)$$

His results are obtained from He to Ne, and we show his total scattering factor for Ne, in which the total charge density is indeed spherical, in Fig. 6.3.

Subsequently, extensive calculations of atomic scattering factors from self consistent field charge distributions for 45 atoms and ions

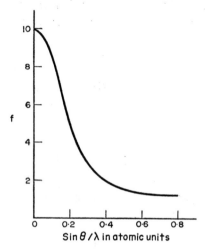

Fɪɢ. 6.3. Scattering factor for Ne atom (cf. Fig. 6.2).

have been carried out by Freeman (1959a). His result for Ne, with a self-consistent field including exchange, is in quite good agreement with that calculated using analytical wave functions and shown in Fig. 6.3.

6.2. Incoherent scattering

(a) *Shape of Compton line in X-ray scattering*

It is well known that when X-rays are scattered by a stationary electron, one obtains from relativistic energy and momentum conservation laws for a photon colliding with an electron, a single modified wavelength (the Compton effect) which is given by

$$\Delta\lambda = \frac{2h}{mc}\sin^2\frac{\phi}{2}$$

where ϕ is the scattering angle and $h/mc = 0.0243$ Å.

However, when X-rays are scattered from an atom, the electrons have a certain velocity or momentum distribution. The incident photon

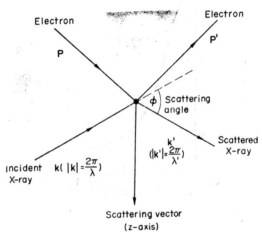

Fig. 6.4. Schematic illustration of Compton scattering, showing interaction of X-ray with an electron of momentum **p** (after Cooper, 1971).

will carry away with it some information about the momentum distribution, through a kind of Doppler effect.

Figure 6.4 illustrates schematically a situation in which an electron has a momentum **p**. If we oversimplify and write down non-relativistic conservation laws, (compare Cooper, 1971) then we have for

(a) Momentum conservation

$$\hbar\mathbf{k}' - \hbar\mathbf{k} = \mathbf{p} - \mathbf{p}'. \tag{6.18}$$

(b) Energy conservation

$$\hbar ck' - \hbar ck = \frac{1}{2m}(p^2 - p'^2).$$ (6.19)

Defining an X-ray scattering vector $\mathbf{s} = \mathbf{k}' - \mathbf{k}$ and assuming that $|s| \gg |k| - |k'|$,[†] the momentum p' can be eliminated to yield

$$\lambda' - \lambda = \frac{2h}{mc}\sin^2\frac{\phi}{2} - \frac{2\lambda\sin\frac{\phi}{2}}{mc}p_z.$$ (6.20)

The second term is seen to lead to a broadening of the Compton line, the z axis being the direction of the scattering vector $\mathbf{s}$.

The modified Compton wavelength, and the Doppler broadening due to the electronic velocity distribution is shown in Fig. 6.5.

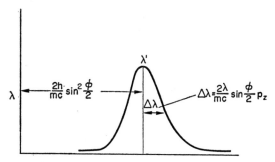

FIG. 6.5. Modified Compton scattering, showing Doppler broadening due to electronic velocity distribution. The figure is a schematic form of the Compton profile in a typical experiment with MoKα radiation $(\lambda = 0.71\ \text{Å})\ \lambda' - \lambda \sim 0.04\ \text{Å}$. The line shape is symmetric with monochromatic incident X-rays (after Cooper, 1971).

The intensity in the line corresponding to momentum p_z will be proportional to the probability of observing that particular component of momentum, that is to

$$\iint P(p_x p_y p_z)\,dp_x\,dp_y.$$ (6.21)

† This assumption is not needed in a proper relativistic determination.

In the case of an isotropic momentum distribution, the intensity of the Compton line reduces to

$$J(q) = \frac{1}{2} \int_q^\infty \frac{I(p)}{p} \, dp \qquad (6.22)$$

where $I(p) \, dp$ is the probability of finding an electron with momentum between p and $p+dp$, discussed at length in section 5.3.

To illustrate the use of this description of the Compton line shape, we will work out immediately the momentum wave function and the, line profile for the hydrogen atom. We note that (6.22) gives the intensity at displacement l from the peak of the line, l being $(2q/c)\lambda \sin(\varphi/2)$.

(i) *Momentum wave function for ground state of hydrogen atom*

We have, in units of the Bohr radius a_0, the ground state wave function

$$\psi(r) = \pi^{-\frac{1}{2}} \exp(-r). \qquad (6.23)$$

It follows from the Fourier transform relation (A5.2.1), that with $\hbar = 1$ that

$$\phi(\mathbf{p}) = \frac{1}{(2\pi)^{\frac{3}{2}}} \int e^{-i\mathbf{p}\cdot\mathbf{r}} \frac{1}{\pi^{\frac{1}{2}}} \exp(-r) \, d\mathbf{r}. \qquad (6.24)$$

If we choose spherical polar coordinates r, θ, ϕ such that θ is the angle between $\mathbf{r}$ and $\mathbf{p}$ then we find that (cf. section 6.1)

$$\phi(\mathbf{p}) = \frac{1}{2^{\frac{1}{2}}\pi} \int_0^\infty e^{-k r^2} \, dr \int_0^\pi e^{-ipr\cos\theta} \sin\theta \, d\theta$$

$$= \frac{2^{\frac{1}{2}}}{\pi p} \int_0^\infty e^{-r} \sin pr . r \, dr$$

$$= \frac{2^{\frac{3}{2}}}{\pi} \frac{1}{(1+p^2)^2} . \qquad (6.25)$$

To make contact with experiment we note that the probability of the electron having momentum vector **p** lying in volume $d\mathbf{p}$ around **p** may be written as

Probability of momentum lying in $d\mathbf{p}$ around **p**

$$= |\phi(\mathbf{p})|^2 \, d\mathbf{p}. \tag{6.26}$$

More specifically, the probability of the electron having momentum of magnitude between p and $p+dp$ is $|\phi(p)|^2 \, 4\pi p^2 \, dp$

$$= I(p) \, dp. \tag{6.27}$$

Then we have the explicit result

$$I(p) = 32p^2/\pi[1+p^2]^4 \tag{6.28}$$

which is plotted in Fig. 6.6 (a).

Again, the large p behaviour of the wave function $\phi(p)$ is determined by the fact that the hydrogen wave function has a finite non-zero slope at $r = 0$.

(ii) *Compton profile for hydrogen atom*

The Compton profile is related to the momentum distribution function $I(p)$ by eqn. (6.22). Using the result (6.28) for $I(p)$ we find the explicit Compton line shape as

$$J(q) = \frac{8}{3\pi(1+q^2)^3} \tag{6.29}$$

which is shown in Fig. 6.6(b).

For other atoms, if we use a determinantal approximation,[†] then it can be shown that $I(p)$ is a sum of contributions $I_{1s}(p)$, $I_{2s}(p)$, etc. It follows directly from the formula for the Compton profile that $J(q)$ is a sum of terms $J_{1s}(q)$, $J_{2s}(q)$, $J_{2p}(q)$, etc. Calculations are available up to K using the wave functions due to Slater, which are closely related to those of Morse, Young and Haurwitz discussed earlier.

Freeman (1959b, c) has also studied the incoherent scattering of X-rays from Ne, Cu^+ Cu, Zn^{+2} and subsequently also from various

[†] See eqn. (7.7) below.

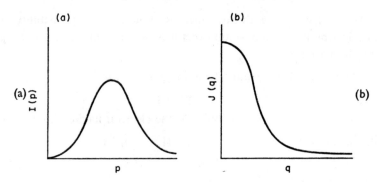

FIG. 6.6. Schematic forms of: (a) Momentum distribution $I(p)$, (b) Compton profile $J(q)$. For atomic hydrogen see eqns. (6.28) and (6.29).

non-spherical charge distributions (e.g. N, N^-,O^-, etc.) using self-consistent field wave functions. The Waller–Hartree (1929) theory was used in the calculations. In concluding this discussion, we draw attention to: (i) the fact that the Thomas–Fermi theory was also used to calculate incoherent scattering (Heisenberg, 1931; see also Pirenne, 1946) before more fundamental work such as Freeman's became available; (ii) the relation between the Waller–Hartree (wH) theory used by Freeman and eqn. (6.22). Apart from involving the one-electron approximation, the result for $J(q)$ follows from the WH theory provided, (a) electronic binding energies are small compared with energy transfer from the photon, and (b) the electron wave functions for the relevant continuum states can be taken as plane waves.

Problems

1. Calculate the atomic scattering factor for the He atom with the Löwdin–Shull open shell wave function (5.32).

2. To contrast with the electron momentum distribution in an atom, calculate the shape of the Compton line $J(q)$ for a non-interacting uniform electron gas at $T = 0$.

(*Hint:* The momentum distribution function $I(p)$ is parabolic, cutting off at the Fermi momentum p_f.)

How would you calculate the momentum distribution of the two core ($1s$) electrons per atom in metallic lithium? (*N.B.* The above free-electron theory would be a useful approximation for the conduction electrons, which in the free atom are the $2s$ electrons.)

CHAPTER 7

Electron exchange and Slater determinants

So far, we have only introduced electron spin into the theory of atoms in the elementary sense of the Pauli exclusion principle. We must now consider the influence of spin in a little more detail. We do so by appeal to the result that the total wave function of a system of fermions must be *antisymmetrical* in the interchange of space and spin coordinates for any pair of particles. To lead up to such a discussion, let us introduce the idea of a spin-orbital by reference to the ground state of the hydrogen atom.

7.1. Spin-orbitals and total wave function for ground state of helium atom

The normalized wave function for the ground state of hydrogen is given, as we have seen, by

$$\psi(r) = \frac{1}{(\pi a_0^3)^{\frac{1}{2}}} \exp\left(\frac{-r}{a_0}\right) : \quad a_0 = \frac{h^2}{4\pi^2 m e^2} . \tag{7.1}$$

To write down the total wave function we must multiply the spatial form (7.1) by either the spin wave function α corresponding to upward spin ($\uparrow$) or β corresponding to downward spin. Thus we have the spin-orbital $\phi(x)$, which has the form $\psi(r)\alpha$ or $\psi(r)\beta$. We have used x here to denote both space and spin coordinates.

Now for the ground state of helium, we have seen that the space wave function $\Psi(\mathbf{r}_1\mathbf{r}_2)$ is symmetrical in the interchange of coordinates r_1 and r_2. The antisymmetric spin function representing the ground state, with opposed spins, has the form

$$\alpha(1)\,\beta(2)-\alpha(2)\,\beta(1), \tag{7.2}$$

1 and 2 denoting the spin coordinates of electrons 1 and 2. Thus, the antisymmetric wave function representing the ground state of the helium atom is given by

$$\Phi(x_1 x_2) = \Psi(r_1 r_2)\,[\alpha(1)\,\beta(2)-\alpha(2)\,\beta(1)]. \tag{7.3}$$

The wave function $\Phi(x_1 x_2 \ldots x_N)$ of a system of N fermions, as we have said, must have this same property of antisymmetry.

7.2. Slater determinants and antisymmetric wave functions

We are evidently interested in setting up here a total wave function built from the personal one-electron wave functions introduced by Hartree. In the case of the helium atom we have seen that our single-particle approximation to the ground-state wave function is

$$\Psi(r_1 r_2) = \psi(r_1)\,\psi(r_2), \tag{7.4}$$

both electrons being in the same space orbital. It is now clear that the total independent-particle wave function

$$\Phi(x_1 x_2)_{\text{independent particle}} = \psi(r_1)\,\psi(r_2)\,[\alpha(1)\,\beta(2)-\alpha(2)\,\beta(1)]. \tag{7.5}$$

It is obvious that the form (7.5) can be rewritten in determinantal form as

$$\Phi(x_1 x_2)_{\text{independent particle}} = \begin{vmatrix} \psi(r_1)\,\alpha(1) & \psi(r_1)\,\beta(1) \\ \psi(r_2)\,\alpha(2) & \psi(r_2)\,\beta(2) \end{vmatrix} \tag{7.6}$$

and we see that the first column contains the spin-orbital $\psi\alpha = \phi_1$, say, while the second contains the other spin-orbital $\psi\beta = \phi_2$. The important point to note is that by interchanging two rows of a determinant we change its sign, and hence obtain the correct antisymmetrical property of the total wave function for fermions.

In the form (7.6) we can effect a generalization to N electrons. The wave function (7.6) is known as a Slater determinant and this approximation to the total wave function is at the heart of all one-electron approximations to the many-body problem. It should be noted that in eqn. (7.6) the rows are labelled by the electron coordinates while the columns correspond to a given spin-orbital.

In general, for a system with N electrons we construct a Slater determinant representing our approximation to the many-body wave function Φ as

$$
\Phi = \begin{vmatrix}
\phi_1(x_1) & \phi_2(x_1) & \dots & \phi_N(x_1) \\
\phi_1(x_2) & \phi_2(x_2) & \dots & \phi_N(x_2) \\
\vdots & & & \\
\phi_1(x_N) & \phi_2(x_N) & \dots & \phi_N(x_N)
\end{vmatrix}. \tag{7.7}
$$

This function, as it stands, is not normalized, even if the ϕ's are.

We can say that whereas Hartree's original theory was based on a product wave function, the Slater determinant represents a properly antisymmetrized product form, obeying the basic symmetry requirement of a system of many fermions.

Whereas in the Hartree theory, we varied the one-electron orbitals making up the product wave function to minimize the energy, we must now vary the ϕ's in the Slater determinant (7.7) to get the best possible approximation to the ground-state energy. When we do this, we shall see a new set of self-consistent field equations emerge, which differ from Hartree's equations by the appearance of non-classical terms in the self-consistent field. These are the exchange terms, arising because of the antisymmetry of the wave function (when electrons with parallel spins are present). Essentially, when we calculate the average value of the term e^2/r_{ij}[†] with a linear combination of product wave functions obtained by expanding eqn. (7.7), we find not only classical electrostatic terms like

$$
\int \psi_1^*(\mathbf{r}_i)\,\psi_2^*(\mathbf{r}_j)\,\frac{e^2}{r_{ij}}\,\psi_1(\mathbf{r}_i)\,\psi_2(\mathbf{r}_j)\,d\mathbf{r}_i\,d\mathbf{r}_j, \tag{7.8}
$$

[†] The Coulomb interaction energy of two electrons i and j separated by distance r_{ij}.

$\psi_1^*(\mathbf{r}_i)\,\psi_1(\mathbf{r}_i)$ being an electron density, but we find also terms of the kind

$$\int \psi_1^*(\mathbf{r}_i)\,\psi_2^*(\mathbf{r}_j)\,\frac{e^2}{r_{ij}}\,\psi_1(\mathbf{r}_j)\,\psi_2(\mathbf{r}_i)\,d\mathbf{r}_i\,d\mathbf{r}_j \qquad (7.9)$$

which evidently has no electrostatic interpretation since $\psi_1^*(\mathbf{r}_i)\,\psi_2(\mathbf{r}_i)$ appears now. These exchange terms, with the repulsive Coulomb interaction, lead to a lowering of the total energy below the Hartree value.

Before outlining the derivation of the equations of the self-consistent field with exchange we will utilize the form of eqn. (7.9) to see how exchange can be included in the Thomas–Fermi theory.

7.3. Inclusion of exchange in Thomas–Fermi theory

The basic Thomas–Fermi theory was set up in Chapter 3 by applying free electron relations locally. In particular the kinetic energy is obtained immediately in this way by noting that the energy per unit volume of N free electrons enclosed in a volume $\mathcal{V}$ is

$$c_k\left(\frac{N}{\mathcal{V}}\right)^{\frac{5}{3}}. \qquad (7.10)$$

The form of this result could be obtained by noting from the Schrödinger equation

$$\nabla^2\psi + \frac{8\pi^2 m}{h^2}[E-V]\psi = 0 \qquad (7.11)$$

that the kinetic energy will depend on the combination h^2/m, and if we assume a dependence on $N/\mathcal{V}$ of the form $(N/\mathcal{V})^\alpha$ then we have, from dimensional analysis:

$$\text{Kinetic energy density} = \text{constant}\left(\frac{N}{\mathcal{V}}\right)^\alpha\left(\frac{h^2}{m}\right)^\beta \qquad (7.12)$$

and we can write the dimensional requirements as

$$ML^2T^{-2}L^{-3} \sim L^{-3\alpha}\left\{\frac{M^2L^4T^{-2}}{M}\right\}^\beta. \qquad (7.13)$$

Hence, it follows immediately by equating powers of mass M on the left- and right-hand sides that $\beta = 1$. Similarly from powers of length L we find

$$\left. \begin{array}{r} -1 = -3\alpha + 4\beta \\ \alpha = \frac{5}{3}, \end{array} \right\} \qquad (7.14)$$

or

and from time T, $\beta = 1$, checking consistency. Needless to say, the constant in eqn. (7.12) cannot be found from dimensional analysis.

The reason we have given this argument is because a similar result can now be found for the exchange energy. This arises from the Coulomb interaction energy e^2/r_{12} between electrons and therefore depends on e^2, rather than $\hbar^2/m$ as with the kinetic energy. Thus, with the same assumptions we can write:

$$\text{Exchange energy density} = \text{constant } (e^2)^{\gamma}\left(\frac{N}{\mathcal{V}}\right)^{\delta} \qquad (7.15)$$

or

$$ML^2T^{-2}\,L^{-3} \sim (LML^2T^{-2})^{\gamma}\,L^{-3\delta}. \qquad (7.16)$$

Immediately we find $\gamma = 1$ by equating powers of M and T while by equating powers of L we find

$$\left. \begin{array}{r} -1 = 3\gamma - 3\delta \\ \delta = 4/3. \end{array} \right\} \qquad (7.17)$$

or

We outline the way the constant in eqn. (7.15) can be calculated in problem (7.5), the final result being:

$$\text{Exchange energy density} = -c_e\left(\frac{N}{\mathcal{V}}\right)^{\frac{4}{3}} \qquad (7.18)$$

where

$$c_e = \frac{3}{4}\,e^2\left(\frac{3}{\pi}\right)^{\frac{1}{3}}.$$

Thus, the total exchange energy A in a model in which we use free electrons relations locally is evidently, with $N/v \rightarrow \varrho(\mathbf{r})$,

$$A = -c_e \int \varrho^{\frac{4}{3}}\,d\mathbf{r}. \qquad (7.19)$$

We now go through the variational argument as before, merely adding this term to the total energy. Its variation is evidently contributing

$$-\tfrac{4}{3} c_e \int \varrho^{\frac{1}{3}} \, \delta\varrho \, d\mathbf{r}. \tag{7.20}$$

Thus, in the Euler equation, we simply add to the term $\tfrac{5}{3} c_k \varrho^{\frac{2}{3}}$ arising from the kinetic energy variation a term $-\tfrac{4}{3} c_e \varrho^{\frac{1}{3}}$ to obtain[†]

$$\tfrac{5}{3} c_k \varrho^{\frac{2}{3}} - \tfrac{4}{3} c_e \varrho^{\frac{1}{3}} + (\mathcal{V} - \mathcal{V}_0)e = 0. \tag{7.21}$$

Thus, the modified relation between density and potential takes the form

$$\varrho = \frac{8\pi}{3h^3} (2me)^{\frac{3}{2}} \left[a + (\mathcal{V} - \mathcal{V}_0 + a^2)^{\frac{1}{2}} \right]^3 \tag{7.22}$$

where $a = (2me^3)^{\frac{1}{2}}/h$. This is the Thomas–Fermi–Dirac (TFD) relation (Dirac, 1930).

We note that only the positive sign has been retained before the square root. Suffice it to say that no physical application has so far been made with the negative sign. If we combine the above relation with Poisson's equation, we obtain a generalization of the Thomas–Fermi equation, namely the TFD equation.

If, by analogy with the way we reduced the Thomas–Fermi equation for atoms to dimensionless form, we write

$$\mathcal{V} - \mathcal{V}_0 + a^2 = \frac{Ze}{r} \phi; \qquad r = bx \tag{7.23}$$

then we find the dimensionless TFD equation

$$\frac{d^2\phi}{dx^2} = x \left[\alpha + \left(\frac{\phi}{x} \right)^{\frac{1}{2}} \right]^3 \tag{7.24}$$

[†] Here $\mathcal{V}_0 e \equiv$ Fermi energy, $-\mathcal{V}e$ is total potential energy. It should be noted, by comparison with the Thomas–Fermi result that we can regard the term proportional to $\varrho^{\frac{1}{3}}$ as an 'exchange' potential—the basis of Dirac–Slater exchange theory.

where

$$\alpha = 6^{\frac{1}{3}}\Big/ 4(\pi Z)^{\frac{2}{3}}.\qquad(7.25)$$

When we look for a solution of this equation, there is one tangential to the x axis, but the point of tangency occurs at a finite value of x and not at infinity as with the Thomas–Fermi equation.

Evaluating the exchange energy (7.19) with the Thomas–Fermi density leads to the approximate result that $A\alpha Z^{\frac{5}{3}}$, which corrects the Thomas–Fermi binding energy $\alpha Z^{\frac{7}{3}}$ by a lower order term for large Z (see problem 7.4).

7.4. Variational derivation of exchange corrections to Hartree theory[†]

If we vary the expectation value of the Hamiltonian with respect to the single-particle wave functions in the Slater determinant, then we are led to Euler equations which are the famous Hartree–Fock equations (cf. Seitz, 1940).

(a) *Euler equations of variation problem: Hartree–Fock equations*

The Hamiltonian for the N electron system may be written in the form

$$H = \sum_i H_i + \frac{1}{2}\sum_{ij}\frac{e^2}{r_{ij}},\qquad (i,j = 1\ldots N)\qquad(7.26)$$

where H_i includes the electronic kinetic energy operators and the electron–nuclear interactions.

We shall assume that we are dealing with a situation in which each single particle state is doubly occupied, the ith state having then the

[†] This section is more advanced than any previous part of the book. It could be omitted by the reader who wishes only to know the physical consequences of including exchange.

wave functions $\psi_i\alpha$ or $\psi_i\beta$, α and β being as usual the spin wave functions. Then we write our approximation to the total wave function ϕ as (compare eqn. (7.7))

$$\phi = (N!)^{-\frac{1}{2}} \det (\Psi_1 \ldots, \Psi_N) \qquad (7.27)$$

or more explicitly, in terms of the permutation operator P,

$$\phi = \frac{1}{(N!)^{\frac{1}{2}}} \sum_P (\pm 1)\, P(\psi_1(\mathbf{r}_1) \ldots \psi_N(\mathbf{r}_N)\, \alpha(\sigma_1)\beta(\sigma_2) \ldots \alpha(\sigma_{N-1})\beta(\sigma_N))$$

$$(7.28)$$

where the space wave functions are equal in pairs, and the plus sign refers to even, and the minus sign to odd, permutations.

We handle the orthogonality and normalization conditions

$$\int \psi_i^*(\mathbf{r})\, \psi_j(\mathbf{r})\, d\mathbf{r} = 0, \quad i \neq j \qquad (7.29)$$

and

$$\int \psi_i^*\psi_i\, d\mathbf{r} = 1 \qquad (7.30)$$

by the customary use of Lagrange multipliers γ_{ij}. Then we must obtain the 'best' single-particle orbitals ψ_i from the variational principle by requiring that

$$\delta\mathcal{E} = \delta \int \phi^* H\phi\, d\mathbf{r}$$

$$= \int \phi^* H\, \delta\phi\, d\mathbf{r} + \int \delta\phi^* H\phi\, d\mathbf{r} = 0. \qquad (7.31)$$

Now from eqn. (7.28) it is readily shown that

$$\delta\phi = \frac{1}{(N!)} \sum_P (\pm 1)\, P\bigg[\sum_i \psi_1(\mathbf{r}_1) \ldots \psi_{i-1}(\mathbf{r}_{i-1})\, \psi_{i+1}(\mathbf{r}_{i+1})$$

$$\ldots \psi_N(\mathbf{r}_N)\, \delta\psi_i(\mathbf{r}_i)\, \alpha_i(\sigma_i) \ldots \alpha_N(\sigma_N)\bigg]. \qquad (7.32)$$

Substituting in eqn. (7.31), summing over spin and using eqns. (7.29) and (7.30) we then find

$$
\sum_i \int \left(\psi_i^*(\mathbf{r}_1) \left[\sum_j \int \psi_j^*(\mathbf{r}_2) H_2 \psi_j(\mathbf{r}_2) \, d\mathbf{r}_2 \right. \right.
$$

$$
+ \frac{1}{2} \sum_{j,k}' e^2 \int \frac{|\psi_j(\mathbf{r}_2)|^2 \, |\psi_k(\mathbf{r}_3)|^2}{r_{23}} \, d\mathbf{r}_2 \, d\mathbf{r}_3
$$

$$
- \frac{1}{2} \sum_{\substack{jk \\ \text{parallel spins}}}' e^2 \int \frac{\psi_j^*(\mathbf{r}_2) \, \psi_k^*(\mathbf{r}_3) \, \psi_j(\mathbf{r}_3) \, \psi_k(\mathbf{r}_2)}{r_{23}} \, d\mathbf{r}_2 \, d\mathbf{r}_3 + H_i
$$

$$
+ e^2 \sum_j \frac{|\psi_j(\mathbf{r}_2)|^2}{r_{12}} \, d\mathbf{r}_2 + \gamma_{ii} \Bigg]
$$

$$
- \left\{ \sum_{\substack{j \\ \text{parallel spins}}} \psi_i^*(\mathbf{r}_1) \left[e^2 \int \frac{\psi_i^*(\mathbf{r}_2) \, \psi_j(\mathbf{r}_2)}{r_{12}} \, d\mathbf{r}_2 \right. \right.
$$

$$
\left. \left. \left. + \int \psi_i^*(\mathbf{r}_2) H_2 \psi_j(\mathbf{r}_2) + \gamma_{ji} \right] \right\} \right) \delta\psi_i(\mathbf{r}_1) \, d\mathbf{r}_1
$$

$$
+ \text{a similar expression in} \quad \delta\psi_i^*(\mathbf{r}_1) = 0. \tag{7.33}
$$

The Lagrange multipliers γ_{ji} are conveniently redefined through

$$
\lambda_{ii} = \gamma_{ii} + \sum_j \int \psi_j^*(\mathbf{r}_2) H_2 \psi_j(\mathbf{r}_2) \, d\mathbf{r}_2
$$

$$
+ \frac{1}{2} \sum_{j,k}' e^2 \int \frac{|\psi_j(\mathbf{r}_2)|^2 \, |\psi_k(\mathbf{r}_3)|^2}{r_{23}} \, d\mathbf{r}_2 \, d\mathbf{r}_3
$$

$$
- \frac{1}{2} \sum_{\substack{j,k \\ \text{parallel spins}}} e^2 \int \psi_i^*(\mathbf{r}_2) \, \psi_k^*(\mathbf{r}_3) \frac{\psi_j(\mathbf{r}_3) \, \psi_k(\mathbf{r}_2)}{r_{23}} \, d\mathbf{r}_2 \, d\mathbf{r}_3 \tag{7.34}
$$

and

$$
\lambda_{ij} = \gamma_{ij} + \sum_j \int \psi_i^*(\mathbf{r}_2) H_2 \psi_j(\mathbf{r}_2) \, d\mathbf{r}_2. \tag{7.35}
$$

At this point we equate the coefficient of $\delta\psi(\mathbf{r}_1)$ to zero, and then we find

$$\left[H_i+\sum_j e^2\int\frac{|\psi_j(\mathbf{r}_2)|^2}{r_{12}}\,d\mathbf{r}_2+\lambda_{ii}\right]\psi_i$$

$$-\sum_{\text{parallel spins}}'\left[e^2\int\frac{\psi_j^*(\mathbf{r}_2)\,\psi_i(\mathbf{r}_2)}{r_{12}}\,d\mathbf{r}_2+\lambda_{ij}\right]\psi_j=0. \qquad (7.36)$$

Taking suitable linear combinations of $\psi_1\ldots\psi_N$ to form new single particle wave functions $\phi_1\ldots\phi_N$, eqn. (7.36) can be brought into diagonal form.

Equations (7.36) represent the Hartree–Fock equations for particles interacting by Coulomb forces.[†] If we neglect the last term, which singles out these electrons with spins parallel to the one under discussion, and omit the term $i=j$ from the sum following H_i, we obtain the Hartree equations. These are simpler to interpret, for they tell us that each electron moves in an effective or average field due to the remaining electrons, calculated in the usual electrostatic manner from the charge distribution $-e\,|\psi_j|^2$ of the jth electron. This was, indeed the way in which these equations were first written down by Hartree.

We see then the connection with the simpler assumption made earlier in Chapter 3. There, the Hartree theory was further simplified, by assuming that each particle moved in the same potential field $V(\mathbf{r})$. However, when the one-electron wave functions have thereby been obtained, we can, of course, calculate the expectation value of H with respect to the determinant of these simpler (and orthogonal), one-electron functions. This 'symmetrized' Hartree approximation is often useful.

On the other hand, aside from symmetrizing the potential, the previous discussion has neglected the 'exchange' terms in eqn. (7.36). These cannot be represented by any local potential $V(r)$, but imply either a velocity or energy dependent potential, or, alternatively, the use of a potential matrix. In spite of this, full use of the Hartree–Fock

[†] For a clear physical discussion of these equations, the reader is referred to reprint 4, by Stater, on p. 215 of this volume (especially p. 218).

equations has been made in calculations on atoms (see Hartree, 1957 for references).

However, an important result can be obtained, without actually solving the Hartree–Fock equations. It concerns the physical interpretation of the one-electron eigenvalues in these equations.

(b) *Koopman's theorem*

The philosophy we shall adopt is as follows. We consider a many-electron wave function of determinantal form, in which the highest occupied ground state orbital is replaced by the lowest unoccupied orbital. Let the eigenvalues of the corresponding one-electron equations be E_N and E_{N+1}. Strictly, making such a promotion would involve changing the lower-lying wave functions, because the field in which these electrons move is altered. This would involve complicated re-calculation of all the one-electron functions in the determinant and no simple result emerges.

Nevertheless it is often true on physical grounds that this rearrangement of the electrons which are not excited is small. If we neglect it, we can calculate the energy of the system with the original ground state determinant and with the excited state determinant. The answer is simply the difference $E_{N+1} - E_N$. Thus, the eigenvalues of the excited orbitals give us directly the excited states of the system: this is, in essence, Koopman's theorem.

Problems

1. Calculate the exchange energy for a uniform electron gas in which all the spins are parallel (ferromagnetic electron gas). At what electron density would this ferromagnetic state become lower in energy in the Hartree–Fock approximation than the paired spin state. (*Hint:* use equations of the form (7.10) and (7.18).)

Obtain a quantitative expression for the Fermi hole when all the spins are parallel.

2. Show that the chemical potential μ for a non-magnetic Fermi gas in the Hartree–Fock approximation is related to the expectation values of the kinetic energy T and potential energy V per particle by

$$\mu = \tfrac{5}{3}\langle T \rangle + \tfrac{4}{3}\langle V \rangle. \tag{P7.1}$$

Hence prove that in equilibrium

$$\mu = \frac{E}{N}$$

where E is the total ground state energy of the electron gas of N particles.

Contrast this result with that for the chemical potential of a non-interacting. Fermi gas.

(*N.B.* Equation (P7.1) is in fact an exact relation for a homogeneous electron gas, and not restricted to the Hartree–Fock approximation.)

3. Assume that the Thomas–Fermi–Dirac equation is solved in a spherical cell of radius R, subject to the condition that the density is flat at R; that is $(\partial \varrho / \partial r)_R = 0$.

What would be the pressure p, at $T = 0$, exerted by the electron gas on the spherical boundary of radius R, assuming an electrically neutral cell?

4. Return to problem (4.1), and develop the total energy there (sum of eigenvalues), to show that, in a Coulomb field (with closed shells only to be discussed), to $0\left(Z^{\frac{1}{3}}\right)$

$$E_{\text{eigenvalue sum}} = \left[-\left(\frac{3}{2}\right)^{\frac{1}{3}} Z^{\frac{7}{3}} + \frac{1}{2} Z^2 - \frac{1}{18}\left(\frac{3}{2}\right)^{\frac{2}{3}} Z^{\frac{5}{3}} \right] \frac{e^2}{a_0} \qquad \text{(P7.2)}$$

(see Ballinger and March, 1955).

The term of order Z^2 takes over into the Thomas–Fermi atom (see Scott, 1952). However, the exchange energy comes in at order $Z^{\frac{5}{3}}$, and, assuming one can approximate this with the Dirac form (7.19), plus the use of the Thomas–Fermi (not Thomas–Fermi–Dirac) density, express the exchange energy in terms of $\phi(x)$ for the Thomas–Fermi atom.

5. In a uniform electron gas of mean density ϱ_0, the density of electrons $\varrho(r)$ at distance r from an electron we choose to 'sit on', is given by (Wigner and Seitz, 1934; Slater, 1951, reprint 4 in this volume)

$$\varrho(r) = \varrho_0 g(r)$$

where

$$g(r) = 1 - \frac{9}{2}\left\{\frac{j_1(k_f r)}{k_f r}\right\}^2. \qquad \text{(P7.3)}$$

Here $j_1(x) = [\sin x - x \cos x]/x^2$ and $\hbar k_f$ is the Fermi momentum.

Give a physical argument to show that the mean potential energy per electron is

$$\frac{V}{N} = \frac{1}{2} e \int \varrho_0 \frac{[g(r) - 1]}{r} d\mathbf{r}. \qquad \text{(P7.4)}$$

Hence, derive eqn. (7.18) explicitly.

CHAPTER 8

Electron–electron correlation

THE treatment of the previous chapter, though still of single-particle form, takes a detailed account of the fact that two electrons with parallel spin tend to avoid one another. The account taken above of this point resides in the fact that we could restate the Pauli principle by saying that two parallel spin electrons could not be found simultaneously at the same point in space. Thus, the probability of two parallel spin electrons, i and j say, being found at a distance r_{ij} from one another, tends to zero as r_{ij} tends to zero. Since such 'correlations' between independent Fermi particles arise from the antisymmetry of the wave

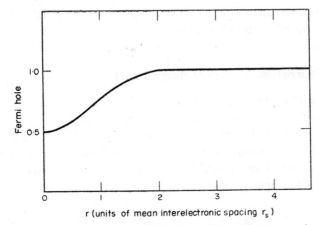

FIG. 8.1. Fermi (or exchange) hole in a uniform electron gas (see, for example, Seitz, 1940; Slater, 1951; reprint 4 in this volume). Oscillations exist at large r but the amplitude is too small to be shown graphically.

function, and hence from the statistics (Fermi) of the particles, they are termed *statistical* correlations.

If we work out the above probability function, using plane waves for the one-electron orbitals, we find the situation shown in Fig. 8.1 (Wigner and Seitz, 1934). This represents the 'Fermi' or 'exchange hole' in a homogeneous system of independent fermions. The probability here is simply a function of $r_{ij} = |\mathbf{r}_i - \mathbf{r}_j|$, since there is no origin singled out in a uniform system. Of course, in an atom, the nucleus provides a natural origin and the exchange hole has a more complex form, as we shall see explicitly below for the Ne atom.

But there are some further physical effects, of course, which arise from the Coulomb repulsions between electrons. Since the statistical correlations already keep parallel spin electrons apart, the situation is not too difficult for these. However, for antiparallel spins, there is no correlation incorporated in the models we have treated so far.[†] This is serious, and we must, at very least, discuss in a qualitative way the consequences of such Coulomb repulsions. The 'hole' around any given electron is therefore partly due to statistical correlations and partly due to Coulomb repulsions. We will give examples below, which, however, we can only discuss in rather general terms, of both exchange and Coulomb holes in light atoms.

8.1. Descriptions in terms of charge density

Because of the complexity of the exchange and Coulomb hole in a non-uniform electronic cloud such as we have in an atom, it is important to enquire whether progress can be made in terms of the charge density. This would develop then the philosophy underlying the Thomas–Fermi theory, which describes the electronic system solely in terms of the density $\varrho(\mathbf{r})$.

[†] In Fig 8.1, the 'Fermi hole' has the value $\frac{1}{2}$ at the origin. This is due to the fact that there is no correlation between antiparallel spins in this Fermi gas model. When such Coulombic repulsion is introduced, the value at the origin is $< \frac{1}{2}$, though the 'hole' itself still contains exactly one electronic charge (see problem 7.5, eqn. (P7.3) for the explicit form of the Fermi hole).

(a) *Generalization of Thomas–Fermi theory to include correlation effects*

This idea was taken up by a number of workers, who proposed to add to the Thomas–Fermi–Dirac energy given in Chapter 7, a term representing the correlation energy as a function of density. A form of the correlation energy obtained from the uniform electron gas was again used, based on approximate calculations by Wigner (1934, 1938) and later workers (see March, Young and Sampanthar, 1967 for a summary). This relation was again used locally, as for kinetic and exchange energies. Thus, if $\varepsilon_c(\varrho)$ is the correlation energy density thereby obtained, we simply add to the Thomas–Fermi–Dirac energy the correlation energy

$$E_{\text{corr}} = \int \varepsilon_c(\varrho)\, d\mathbf{r} \qquad (8.1)$$

and repeat the minimization. The approximate form for $\varepsilon_c(\varrho)$ according to Wigner (1938) is, in atomic units,

$$\varepsilon_c(\varrho) = -\frac{0{\cdot}056\varrho^{\frac{4}{3}}}{0{\cdot}079 + \varrho^{\frac{1}{3}}}. \qquad (8.2)$$

If we carry through the minimization, we get a new Euler equation, incorporating correlation effects (an explicit example of this procedure, in a different context, is the work of Smith, 1969, on the electron density near a metal surface).

The objection to this procedure is that the kinetic energy is approximate already, being correct only for slowly varying potentials. It might well be questioned whether corrections for correlations would be significant, in view of the errors involved in the Thomas–Fermi method itself.

Though we cannot go into detail here, the work of Hohenberg and Kohn (1964) has shown that the philosophy behind the Thomas–Fermi theory is completely justified even in a many-electron problem. The ground energy is in fact a unique functional of the particle density. The major problem, of course, is to find the functional and this is equivalent to solving the many-body problem. However, this extension of the Thomas–Fermi philosophy is very attractive and will undoubtedly gain increasing prominence in the future.

That an exact form of the Thomas–Fermi method is indeed possible for particles moving *independently* in a common potential $V(\mathbf{r})$ is explicitly demonstrated, to all orders in a perturbation expansion in $V(\mathbf{r})$, in the work of March and Murray (1961) and Stoddart and March (1967). The formal theory of Hohenberg and Kohn (1964) shows that such an approach still has a fundamental foundation when extended to *interacting* electrons.

While dealing with descriptions in terms of the charge density, we take the opportunity to draw attention to an exact, if limited, relation in atoms between the charge density and its gradient, both evaluated at the nucleus.

(b) *Relation between charge density and its gradient at nucleus*

Kato (1957) has proved the important result that the wave function Ψ for a system of N electrons[†] satisfies the condition

$$\left. \frac{\partial \overline{\Psi}}{\partial r_n} \right|_{r_n=0} = -Z[\overline{\Psi}(\mathbf{r}_1 \mathbf{r}_2 \ldots \mathbf{r}_N)]_{\mathbf{r}_n=0} \qquad (8.3)$$

where

$$\overline{\Psi} = \frac{1}{4\pi} \int_{\omega_n} \Psi(\mathbf{r}_1 \mathbf{r}_2 \ldots \mathbf{r}_N)\, d\omega_n \qquad (8.4)$$

represents the average of Ψ over a sphere of radius r_n: that is an average over polar angles θ and ϕ for the nth electron is carried out. Here it is assumed that the coordinate system has its origin at the nucleus we are considering.

[†] The Hamiltonian is assumed spin-independent in deriving this result.

As a corollary to this result, it has been shown by Steiner (1963) that a relation can be derived between the charge density and its derivative at the nucleus. A sketch of his treatment is given in Appendix 8.1.

Let us consider the very elementary example of the hydrogen atom. Then the ground state wave function (cf. eqn. (2.1))

$$\psi = \frac{1}{(\pi a_0^3)^{\frac{1}{2}}} \exp\left(-\frac{r}{a_0}\right) \qquad (8.5)$$

gives a charge density

$$\varrho(r) = \psi^2 = \frac{1}{\pi a_0^3} \exp\left(-\frac{2r}{a_0}\right). \qquad (8.6)$$

The formula we prove in Appendix 8.1 is

$$\left(\frac{\partial \varrho}{\partial r}\right)_{r=0} = -\frac{2\varrho(0)}{a_0} \qquad (8.7)$$

for $Z = 1$, and from the above result (8.6) we see that

$$\varrho(0) = (\pi a_0^3)^{-1} \qquad (8.8)$$

while

$$\left(\frac{\partial \varrho}{\partial r}\right)_{r=0} = -\frac{2}{\pi a_0^4} \qquad (8.9)$$

which obviously agrees with the theorem. Its importance, of course, is that it holds exactly in an atom with many interacting electrons and sometimes affords a useful criterion to test approximate many-electron wave functions.

8.2. Exchange and correlation holes in atoms

(a) *Helium–like ions*

We shall focus again on the pair correlation between electrons in this section. In the ground state of the helium atom, there are no exchange terms, since the electrons have antiparallel spins.

8*

The Coulomb repulsion between electrons will obviously be reflected in the mean value of $\langle r_{12} \rangle$, r_{12} being the interelectronic distance as usual. In particular, we must expect that $\langle r_{12} \rangle$ will increase as we introduce electronic repulsion into an uncorrelated wave function.

To see how this increase comes about, it is useful to introduce a distribution function $f(r_{12})$ which is normalized such that

$$\int_0^\infty f(r_{12}) \, dr_{12} = 1. \tag{8.10}$$

This distribution function will show us the effect of electron–electron correlations rather directly if we calculate if from the best available wave function and compare it with an uncorrelated wave function, which we shall choose as the 'effective nuclear charge' function (2.17) with $Z' = Z - \frac{5}{16} = \frac{27}{16}$ for the helium atom.

Explicitly $f(r_{12})$ is related to the space wave function $\Psi(\mathbf{r}_1\mathbf{r}_2)$ by

$$f(r_{12}) \, dr_{12} = \int \Psi^2(\mathbf{r}_1\mathbf{r}_2) \, d\mathbf{r}_1 \, d\mathbf{r}_2 \tag{8.11}$$

where, however, the integrations are performed over all positions of the two electrons such that the interelectronic distance lies between r_{12} and $r_{12} + dr_{12}$.

From the 'effective nuclear charge' wave function, we find for f the result

$$f(x) = \frac{Z'^3}{6} (3x^2 + 6Z'x^3 + 4Z'^2x^4) \, e^{-2Z'x} \tag{8.12}$$

and this is shown in curve 1 of Fig. 8.2.

Though we have not discussed very refined wave functions for He earlier, a very good approximation to the ground-state eigenfunction of the Hamiltonian operator (2.6) can be written in the form

$$\Psi(\mathbf{r}_1\mathbf{r}_2) = \Sigma C_{lmn}(r_1 + r_2)^l \, (r_1 - r_2)^{2m} \, r_{12}^n e^{-Z'(r_1 + r_2)} \tag{8.13}$$

Including six terms in this expansion, the coefficients of which were determined variationally by Hylleraas (1929) it is still possible to carry through the calculation of $f(x)$ (Coulson and Neilson, 1961) and their result is shown in curve 2 of Fig. 8.2. This must be very near to the exact result for He. Curve 3 shows the self-consistent field theory, using the analytic wave function of Roothaan, Sachs and Weiss (1960) rather than the numerical Hartree wave function of Wilson and Lindsay which we referred to in Chapter 2.

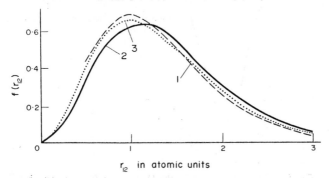

FIG. 8.2. $f(r_{12})$ for He atom (after Coulson and Neilson, 1961). Curve 1. Result of eqn. (8.12). Curve 2. Result of Hylleraas wave function (8.13). Curve 3. Result of self-consistent field theory.

There is a marked broadening out of $f(x)$ as we go from curve 1 to curve 2, and the mean value $\langle r_{12} \rangle$ increases from $1·296a_0$ to $1·420a_0$. The corresponding change in the interelectronic repulsion energy $\langle e^2/r_{12} \rangle$ is from $1·055(e^2/a_0)$ to $0·946(e^2/a_0)$, which is about 3 eV. Though this is a substantial reduction, the kinetic energy is thereby raised, and from the virial theorem only half of this energy represents a gain in total energy. This is a general consequence of correlating electronic motions: we lose on the kinetic energy term but regain more potential energy.

The correlation hole, defined as

$$\Delta f(x) = f_{\text{exact}} - f_{\text{Hartree-Fock}} \tag{8.14}$$

has the form shown in Fig. 8.3 where we see that, in He, the chance of two electrons lying anywhere within a Bohr radius of each other is less than it would be without electron correlation. Correspondingly,

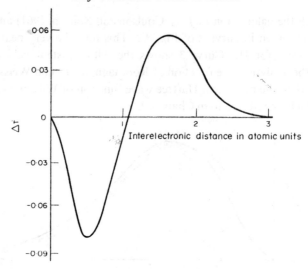

FIG. 8.3. Correlation hole for the He atom, defined in eqn. (8.14).

the probability that the electrons are separated by more than $\sim a_0$ is greater.

Figure 8.4 shows results for the heavier two-electron ions, Li^+ ... ... Ne^{8+}, for comparison with the He atom results (Lester and Krauss, 1964). If, as in Fig. 8.4, we plot Δf against Zr_{12}, the curves vary only slowly with Z, and hardly change shape at all.

When electrons with parallel spins are involved, we must consider exchange and correlation together in a full treatment, as we have seen. We shall next discuss the exchange hole in neon, as studied by Maslen (1956) and we shall refer briefly to calculations on Be in which both exchange and correlation effects have been included.

(b) *Fermi hole in Ne and Coulomb hole in Be*

Maslen has studied the exchange charge density, which Slater (1951; see reprint 4 in this volume) has focused on, for neon and argon. In this way, the size and shape of the Fermi hole can be studied, and

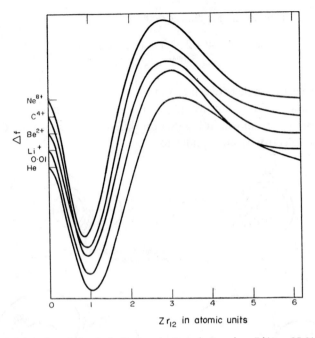

FIG. 8.4. Correlation holes for heavier two-electron ions $Li^+ \ldots Ne^{8+}$ (after Lester and Krauss, 1964).

the accuracy of the Slater approximation of a spherical exchange charge density centred on the electron assessed (see reprint 4, p. 215).

To avoid the complexity of working with numerical self-consistent field wave functions, Maslen used Slater orbitals to calculate the average exchange hole (Slater, 1951; reprint 4, eqn. (6), p. 222)

$$-\left[\sum_{\substack{j,\,k=1 \\ \|\text{spins}}}^{N} \Sigma \psi_j^*(1)\,\psi_k^*(2)\,\psi_k(1)\,\psi_j(2)\right] \Bigg/ \left[\sum_{j=1}^{N} \psi_j^*(1)\,\psi_j(1)\right]. \quad (8.15)$$

For Ne he adopted the normalized Slater functions (Slater, 1930; see also Appendix 6.1)

$$\left. \begin{array}{c} \psi(1s) = 17\!\cdot\!04e^{-9\cdot7r}, \quad \psi(2s) = 4\!\cdot\!67re^{-2\cdot9r} \\ \psi(2p_x) = 8\!\cdot\!08xe^{-2\cdot9r} \end{array} \right\} \quad (8.16)$$

which yields an exchange charge density

$$\frac{[290{\cdot}4e^{-9{\cdot}7(r_1+r_2)}+21{\cdot}75r_1r_2(1+3\cos\gamma)e^{-2{\cdot}9(r_1+r_2)}]^2}{290{\cdot}4e^{-19{\cdot}4r_1}+87{\cdot}0r_1^2e^{-5{\cdot}8r_1}} \qquad (8.17)$$

where γ is the angle between $\mathbf{r}_1$ and $\mathbf{r}_2$.

Contours of constant values of the exchange charge density, in units of $+e$ are shown in Fig. 8.5 for different choices of the distance r_1 of the electron from the nucleus.

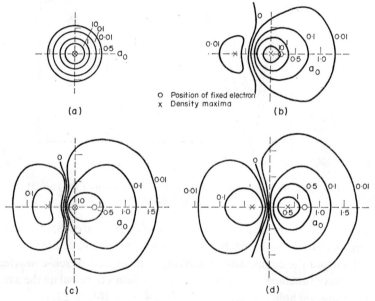

FIG. 8.5. Exchange charge density contours for Ne atom, for various values of r_1 (a) $r_1 = 0$, (b) $0{\cdot}2a_0$, (c) $0{\cdot}4a_0$, (d) $0{\cdot}7a_0$ (after Maslen, 1956).

The form of the exchange charge density is evidently fairly complex and spherical approximations are quite rough.

Smith (1971) has studied the exchange and correlation hole in Be using Hartree–Fock calculations and comparing with variational calculations of Weiss. (1961) which transcend the Hartree–Fock approximation and therefore include correlation effects. The Coulomb hole, analogous to that shown to He in Fig. 8.2, has thus been studied also for the ground state of Be. X-ray scattering in fact can give quite direct information on the pair correlation function between electrons in atoms but that leads us to a deeper formulation of X-ray scattering than that presented in Chapter 6 and is outside the scope of this book, (see, Jones and March, 1973).

(c) *Binding energies of atoms by* $1/Z$ *expansion*

The theory of atomic binding energies has progressed along two directions, which have hitherto been largely independent developments. Suppose we consider the total electron energy of an ion; with nuclear charge Ze and N electrons. Then when Z is large, we can write down the total energy in the form

$$E(Z, N) = Z^2\left[\varepsilon_0 + \frac{1}{Z}\varepsilon_1 + \frac{1}{Z^2}\varepsilon_2 + \frac{1}{Z^3}\varepsilon_3 + \ldots\right] \qquad (8.18)$$

where all the N dependence is contained in the coefficients ε_n in (8.18), and convergence is guaranteed for sufficiently large Z, as shown by Kato (1951).

Now these coefficients are available, in low order, through the work of a number of groups, for the $N = 2$ to 10 ground states, and their

TABLE 8.1 ENERGY COEFFICIENTS IN EQUATION (8.18) FOR Z^{-1} THEORY. (Energies are in atomic units; references in March and White, 1972).

N	Configura-tion		$-\varepsilon_0$	ε_1	$-\varepsilon_2$	ε_3
2	^1S	$1s^2$	1	0·625	0·15766	0·008699
3	^2S	$2s$	1·125	1·0228	0·40837	−0·0230
4	^1S	$2s^2$	1·25	1·5593	0·8819	
5	^2P	$2p$	1·375	2·3275	1·8567	
6	^3P	$2p^2$	1·50	3·2589	3·2880	
7	^4S	$2p^3$	1·625	4·3535	5·2640	
8	^3P	$2p^4$	1·75	5·6619	8·1319	
9	^2P	$2p^5$	1·875	7·1343	11·7551	
10	^1S	$2p^6$	2·00	8·7708	16·2729	

results are summarized in Table 8.1. Some comments on the way ε_0 and ε_1 are found are made at the end of this section.

The other approach is to ask for the binding energy of a heavy neutral atom $E(Z, Z)$ as Z becomes very large. As we saw in Chapter 4 (see also problem 7.4), the result can be written in the form

$$E(Z, Z) = \frac{e^2}{a_0} \left[-0{\cdot}77 Z^{\frac{7}{3}} + \frac{1}{2} Z^2 + 0\left(Z^{\frac{5}{3}}\right) \right]. \qquad (8.19)$$

It is clear from the Thomas–Fermi derivation of the leading term in eqn. (8.19) that we have involved in the coefficient of $Z^{\frac{7}{3}}$ the screening of the nuclear Coulomb field by the Z electrons. Thus, in the form (8.19) we must have summed sub-series to all orders in Z of eqn. (8.18), in such a way that we have used the asymptotic forms of the coefficients $\varepsilon_n(N)$ for large N, in the limit $N \to Z$.

From eqn. (8.19), it is obviously not possible to unscramble the terms in the $O\left(Z^{\frac{7}{3}}\right)$ contribution, for instance, to compare with eqn. (8.18).

But we shall show below that this can be done by using the Thomas–Fermi theory for positive ions (compare Chapter 4) to set up the theory of $E(Z, N)$ directly. This theory is only valid asymptotically, but, as we shall show, is useful for giving an asymptote into which the results of $\varepsilon_n(N)$ shown for $N \leqslant 10$ in Table 8.1 must eventually run. (It should be noted that the $1/Z$ expansion includes electron correlation, however).

We have from eqn. (4.18) the formal result

$$E(Z, N) = Z^2 \left[Z^{\frac{1}{3}} f\left(\frac{N}{Z}\right) \right] \frac{e^2}{a_0}, \qquad (8.20)$$

the function $f(N/Z)$ simply expressing the fact that $\phi'(0)$ and x_0 in eqn. (4.18) depend solely on N/Z. The form of $f(N/Z)$ is given in Fig. 4.1.

This expression (8.20) must now be compared with the form (8.18) when we regard both N and Z as large.

It is straightforward from the hydrogen atom theory of Chapter 2 (cf. March and White, 1972) to show that

$$\varepsilon_0(N) \propto N^{\frac{1}{3}} \qquad (8.21)$$

as N becomes large. We can write this trivially as proportional to $Z^{\frac{1}{3}}(N/Z)^{\frac{1}{3}}$ and similarly it is clear that

$$\varepsilon_1(N) \propto N^{\frac{4}{3}} \tag{8.22}$$

or

$$\frac{1}{Z}\varepsilon_1(N) \propto \left(\frac{N}{Z}\right)^{\frac{4}{3}} Z^{\frac{1}{3}}. \tag{8.23}$$

In general, we expect therefore

$$\frac{1}{Z^n}\varepsilon_n(N) \propto Z^{\frac{1}{3}}\left[\frac{N}{Z}\right]^{n+\frac{1}{3}} \tag{8.24}$$

Representation of coefficients

This suggests that one should use the coefficients $\varepsilon_0(N)$, $\varepsilon_1(N)$ etc., which are known approximately numerically for $N \leqslant 10$ from Table 8.1, to plot $\ln \varepsilon_n(N)$ against $\ln N$. One might then expect, in this way, to verify eqn. (8.24) for large N. But the results for $n = 0$ give, from the above ln–ln plot, an exponent near to $\frac{2}{5}$, from the values up to $N = 10$ known from Table 8.1. In fact, it turns out that N is not large enough in the range 2 to 90 even to yield the characteristic large N dependence of $N^{\frac{1}{3}}$ for $n = 0$. This is surely related to Foldy's result, recorded in Chapter 4, that, in this range of atomic number, the binding energy of neutral atoms goes like $Z^{\frac{12}{5}}$.

However, by investigation of the exact asymptotic form of $\varepsilon_0(N)$, March and White (1972) have shown that this can be predicted from a least squares fit of $\varepsilon_0(N)$ versus $N^{\frac{1}{3}}$, the results being shown in Fig. 8.6. Using the same least squares method for ε_1 and ε_2 on the results of Table 8.1 they find the *approximate* asymptotic forms (cf. eqn. (8.24))

$$\varepsilon_1 = 0{\cdot}661N^{\frac{4}{3}} - 0{\cdot}611N^{\frac{1}{3}} + \ldots \tag{8.25}$$

and

$$\varepsilon_2 = -0{\cdot}142N^{\frac{7}{3}} + 0{\cdot}158N^2 + \ldots \tag{8.26}$$

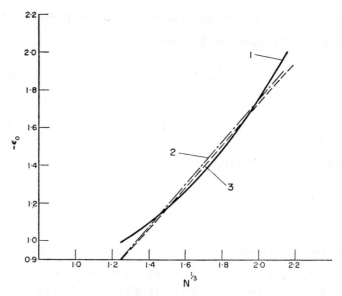

FIG. 8.6. Coefficient $\varepsilon_0(N)$ versus N in $1/Z$ expansion. Curve 1. Exact results. Curve 2. Least squares fit. Curve 3. Exact asymptotic form (March and White, 1972).

This then establishes, at least in principle, the relation between the Thomas–Fermi and the $1/Z$ theories.

Having discussed total binding energies we want to make a few more general remarks about the $1/Z$ expansion.

In providing motivation for such an expansion, Layzer (1959) drew attention to some observed regularities in atomic properties, for the interpretation of which numerical Hartree–Fock studies are not very illuminating. Thus:

(i) The square root of the ionization potential varies linearly with Z in a sequence of homologous ions (i.e. ions that have equal numbers of electrons and are in corresponding states).

(ii) The difference in energy between two spectroscopic terms in the same configuration varies linearly with Z along an isoelectronic sequence.

The departures found from (i) and (ii) are surprisingly small and they have played an important role in experimental studies of the spectra of highly ionized atoms. As remarked already, Hartree–Fock theory is not well suited to study such Z dependence.

Without going into detail about the results of the $1/Z$ method, it should be said that the eigenvalues of the Hamiltonian $H(N, Z)$, for an ion of nuclear charge Z and N electrons, which take the form (8.18) have the property that

  (i) The coefficient ε_0 is given by the hydrogenic levels as

$$\varepsilon_0 = -\sum_{i=1}^{N} \frac{1}{2n_i^2} \tag{8.27}$$

  (ii) The coefficients ε_1 are eigenvalues of a certain matrix formed
       from

$$\sum_{i<j} e^2/r_{ij}$$

where the matrices, however, are to be evaluated in a representation based on hydrogenic radial wave functions. Such calculations have been carried out to obtain ε_1 as given in Table 8.1.

We should note finally that we can infer from the $1/Z$ expansion that the ionization potential has the form

$$\frac{(Z-\sigma)^2}{2n^2} + C + O(Z^{-1}), \tag{8.28}$$

though it is not clear that the third term is small compared with the sum of the first two, when Z is not much greater then N.

Again, it follows that the difference in energy between two terms belonging to the same set of radial quantum numbers has the form

$$AZ + B + O(Z^{-1}). \tag{8.29}$$

These forms (8.28) and (8.29) go a long way towards understanding (i) and (ii) above. For a detailed comparison between the results of $1/Z$ theory and experiment, reference may be made of Edlen (1971).

8.3. Collective effects

The one-electron theories described in the earlier chapters have had great success in dealing with many of the physical properties of atoms. But, as we saw, they replace the Coulombic repulsion term $\sum_{i<j} e^2/r_{ij}$ in the many-body Hamiltonian by some average field, and thereby fail to treat the detailed correlations in the electronic motion induced by the electrostatic repulsions. As we stressed only statistical correlations are included in the Hartree–Fock theory, via the Fermi or exchange hole which is a region deficient of electrons with spins parallel to that of the electron the hole surrounds. Of course, because of the Coulomb repulsion, antiparallel spin electrons are also repelled and this Coulomb hole is not correctly described by the one-electron theories. For light atoms, this Coulomb hole was discussed by *ad hoc* methods in section 8.2; we now want to turn to the case of a many-electron system, and enquire what qualitative effects these electron-electron interactions can cause.

In a uniform gas of electrons, we shall show below that the long-range Coulomb interactions lead to organized collective modes of the electron gas as a whole: with the plasma frequency

$$\omega_p = \left(\frac{4\pi\varrho_0 e^2}{m}\right)^{\frac{1}{2}} \tag{8.30}$$

where ϱ_0 is the electron density.

(a) *Plasma oscillations in uniform electron gas*

We assume a uniform background density of positive charge $e\varrho_0$ and denote by $\varrho(\mathbf{r}t)$ the density of electrons at position $\mathbf{r}$ at time t. The excess positive charge is clearly given by $e(\varrho_0 - \varrho)$ and hence from Maxwell's equations we can write

$$\text{div } \mathcal{E} = 4\pi e(\varrho_0 - \varrho) \tag{8.31}$$

where $\mathcal{E}$ is the electric field.

Now we displace the electron gas by $\mathbf{x}$ to give a current density $\varrho\dot{\mathbf{x}}$ and we have the equation of continuity

$$\text{div}\,(\varrho\dot{\mathbf{x}}) = -\frac{\partial\varrho}{\partial t}. \tag{8.32}$$

If we assume the displacement $\mathbf{x}$ to be small, then the plasma oscillations are small in amplitude and we may write (8.32) as

$$\varrho_0\,\text{div}\,\dot{\mathbf{x}} = -\frac{\partial\varrho}{\partial t}. \tag{8.33}$$

This equation can immediately be integrated to give

$$\varrho_0 - \varrho = \varrho_0\,\text{div}\,\mathbf{x} \tag{8.34}$$

where we have used the boundary condition that $\varrho = \varrho_0$ when $\mathbf{x} = 0$. Thus we have the result

$$\text{div}\,\mathcal{E} = 4\pi e\varrho_0\,\text{div}\,\mathbf{x} \tag{8.35}$$

when we substitute eqn. (8.34) into (8.31). Hence, we can write

$$\mathcal{E} = 4\pi e\varrho_0\mathbf{x} \tag{8.36}$$

which evidently satisfies the correct boundary condition that $\mathcal{E} = 0$ when $\mathbf{x} = 0$. If this result is now combined with the Newtonian equation of motion for an electron in an electric field $\mathcal{E}$, namely

$$m\ddot{\mathbf{x}} = -e\mathcal{E} \tag{8.37}$$

then we have

$$m\mathbf{x} + 4\pi e^2\varrho_0\mathbf{x} = 0. \tag{8.38}$$

This immediately shows us that the electron gas will oscillate, with an angular frequency ω_p, the plasma frequency, given by eqn. (8.30) above. This argument is useful in metals, where ω_p is a very high frequency, typically 10^{16} per sec for the conduction electrons in a nearly-free electron metal. However, the electronic cloud in an atom is, of course, very inhomogeneous. Nevertheless, a start can be made on the theory of collective motions in such an electronic cloud. This, of course, is then in marked contrast to the one-electron theory of the self-consistent-field methods. Though this area is rather undeveloped, we shall present some evidence below that there are collective effects in atoms, though the collective motions are nothing like so sharply defined as in the uniform electron gas discussed above, where, for example, direct experimental verification of the excitation of quanta of the plasma oscillations is found by firing fast electrons into thin metal films.

(b) *Bloch's hydrodynamic theory*

Bloch (1933) considered the way in which the above argument for collective modes in a uniform electron gas might be modified in a spatially varying charge cloud. As examples of his work, one might cite the conduction electrons in a metal, in which the electron density obviously piles up round the positive ions (see March and Tosi, 1972, for a quantum theory of collective modes in such a system), or in the present context, the electronic charge cloud in a heavy atom.

Bloch, in his pioneering work, considered the possible modes of oscillation of such an inhomogeneous electron gas and though no realistic application of this theory has been made to atoms, we shall summarize the basis of his work. He obtained the equation of motion of the gas from a variation principle, by writing

$$\delta \int_{t_1}^{t_2} L \, dt = 0 \qquad (8.39)$$

with the Lagrangian L given by

$$L = m \int \varrho \frac{\partial w}{\partial t} \, d\mathbf{r} - H. \qquad (8.40)$$

Here ϱ is, as usual, the number of electrons per unit volume, while w is the velocity potential. This is related to the velocity $\mathbf{v}$ by

$$\mathbf{v} = -\operatorname{grad} w. \qquad (8.41)$$

H is the energy of the electron gas in motion and if we use the Thomas–Fermi–Dirac theory of Chapter 7 we may write

$$H = \tfrac{1}{2}m \int \varrho(\text{grad } w)^2 \, d\mathbf{r} - \int (V_N + \tfrac{1}{2}V_e + V_{\text{ext}})e\varrho \, d\mathbf{r}$$

$$+ c_k \int \varrho^{\frac{5}{3}} \, d\mathbf{r} - c_e \int \varrho^{\frac{4}{3}} \, d\mathbf{r}. \tag{8.42}$$

V_N and V_e are respectively the potentials created by the nuclei and the electronic cloud ϱ, while V_{ext} denotes an applied external time-dependent potential. Using the variational principle, one then finds (Bloch, 1933), with $V = V_N + V_e$,

$$m\frac{\partial w}{\partial t} = \frac{1}{2} m(\text{grad } w)^2 - (V + V_{\text{ext}})e + \frac{5}{3} c_k \varrho^{\frac{2}{3}} - \frac{4}{3} c_e \varrho^{\frac{1}{3}} \tag{8.43}$$

and the equation of continuity (8.32) reads

$$\frac{\partial \varrho}{\partial t} = \text{div} (\varrho \text{ grad } w). \tag{8.44}$$

If one uses the expression for the pressure of the electron gas as a function of the electron density, which in the Thomas–Fermi–Dirac approximation has the free-electron form (see problems 8.4 and 7.3)

$$p = \tfrac{2}{3}c_k\varrho^{\frac{5}{3}} - \tfrac{1}{3}c_e\varrho^{\frac{4}{3}} \tag{8.45}$$

then the equation of motion (8.43) can be rewritten as

$$m\frac{\partial w}{\partial t} = \frac{1}{2} m(\text{grad } w)^2 - (V + V_{\text{ext}})e + \int \frac{dp}{\varrho}. \tag{8.46}$$

This equation, together with the equation of continuity, form the basis of Bloch's hydrodynamic theory.

These equations can now be used, by writing

$$\varrho(\mathbf{r}, t) = \varrho_0(\mathbf{r}) + \varrho_1(\mathbf{r}, t) \tag{8.47}$$

and treating ϱ_1 as small, to discuss the eigenmodes of an inhomogeneous electron gas. In particular, Jensen (1937) has considered the relatively crude model

$$\left.\begin{array}{ll} \varrho_0 = Z(\tfrac{4}{3}\pi R^3)^{-1} & \text{for} \quad r < R \\ \varrho_0 = 0 & \quad r > R \end{array}\right\} \qquad (8.48)$$

and has found the eigenmodes.

We shall not go into the details further here, except to say that, linearizing eqns. (8.44) and (8.46) in ϱ_1, one can combine the two equations into an equation for $\partial^2\varrho_1(\mathbf{r}t)/\partial t^2$ which has the structure

$$\frac{\partial^2\varrho_1(\mathbf{r}t)}{\partial t^2} = -\omega_p^2(\mathbf{r})\,\varrho_1(\mathbf{r}t) + \text{terms involving gradients of } \varrho_0 \qquad (8.49)$$

where $\omega_p^2(\mathbf{r})$ is defined in eqn. (8.50) below. It is an equation having the general form (8.49) which essentially Jensen has solved for the crude model density (8.48). But at this stage, let us enquire how collective effects can manifest themselves in atomic properties.

(c) *Response of atoms to applied fields, and collective resonances*

In an inhomogeneous system like an atom, we expect, and Bloch's theory above bears it out, that the 'local plasma frequency'

$$\omega_p(\mathbf{r}) = \left\{\frac{4\pi\varrho_0(\mathbf{r})e^2}{m}\right\}^{\frac{1}{2}} \qquad (8.50)$$

plays a role in any collective behaviour, $\varrho_0(\mathbf{r})$ being as usual the ground-state electron density in the atom, which we have discussed in some detail earlier.

We shall confine ourselves here to the consideration of one type of experimental observation which bears on collective behaviour, namely atomic photoabsorption cross-sections in their gross features. For the finer detail, individual cases must be considered and we shall mention one of these of particular interest in ending this section: the collective effects in the $(4d)^{10}$ shell of Xe.

We adopt the Thomas–Fermi model and thereby we shall find a universal photoabsorption cross-section emerging (Brandt and Lundqvist, 1964; Brandt, Eder and Lundqvist, 1967); just as for many other properties of the Thomas–Fermi atom, as we have seen, there is simple scaling with atomic number. It might be expected that such a universal curve can account for the average behaviour of the absorption cross-sections of atoms at frequencies beyond those reflecting the details of atomic binding, yet lower than those of characteristic X-ray absorption edges.

(i) *Photoabsorption cross-section*

Let us consider an atom in an external field with a definite angular frequency ω, and with a wavelength long relative to atomic dimensions. Then the photoabsorption cross-section, i.e. the photoextinction coefficient per atom can be expressed in the form

$$\sigma(\omega) = (2\pi^2 e^2/mc)\,g(\omega). \qquad (8.51)$$

The function $g(\omega)$ on which this formula focusses is, in technical language, the differential oscillator strength distribution or the spectral function. This quantity contains basic information about the dynamic response of atomic systems to external fields, and the function $g(\omega)$, which can be thought of as giving us, when multiplied by $d\omega$, the number of collective modes in the frequency range between ω and $\omega + d\omega$ satisfies then (compare eqn. (8.60) below)

$$\int g(\omega)\,d\omega = N, \qquad (8.52)$$

where N is the number of electrons in the system under consideration.

This spectral function is related to the atomic polarizability $\alpha(\omega)$ and it can be shown (see Brandt and Lundqvist) that $g(\omega)$ is related to the polarizability α through

$$g(\omega) = (2m/\pi e^2)\omega \, \mathrm{Im}\, \alpha(\omega). \qquad (8.53)$$

9*

If we can neglect the gradient terms in the ground state electron density then it can be shown that $\alpha(\omega)$ can be written in the form[†]

$$\alpha(\omega) = -(4\pi e^2/\omega^2) \int_0^\infty r\,dr \int_0^\infty r'\,dr' \int_0^\infty dq \varrho_0(r')\,\phi(qr)\,\varepsilon^{-1}(q, \omega, \varrho_0(r))$$

$$(8.54)$$

In this expression, $\varrho_0(r)$ is as usual the ground state electron density while

$$\phi(qr) = (2/\pi)^{\frac{1}{2}} \sin qr \qquad (8.55)$$

where q is the wave number. The quantity ε is a generalized dielectric constant; it depends on wave number q, frequency ω and the local electron density $\varrho_0(r)$. Combining the two equations above, one finds then

$$g(\omega) = (8\pi/\omega) \int r\,dr \int r'\,dr'\, \varrho_0(r')\,\phi(q_0 r)\,\phi(q_0 r') \left| \frac{d\varepsilon}{dq} \right|_{q_0}^{-1}. \quad (8.56)$$

Here q_0 is the solution of the local dispersion relation

$$\varepsilon(q_0, \omega, \varrho_0(r)) = 0 \qquad (8.57)$$

and the integrations cover those regions where $q_0(\omega, r)$ is real.

We must now specify the local response of the electrons. The simplest assumption which has some of the main features of the problem is to take the Drude form

$$\varepsilon(\mathbf{r}) = 1 - \{\omega_p^2(\mathbf{r})\}/\omega^2 \qquad (8.58)$$

with neglect of spatial dispersion: i.e. with no q dependence. One then finds the atomic polarizability as

$$\alpha(\omega) = \frac{e^2}{m} \int d\mathbf{r}\, \frac{\varrho_0(\mathbf{r})}{\omega_p^2(\mathbf{r}) - \omega^2} \qquad (8.59)$$

[†] The reader may pass immediately to eqn. (8.59) if he is unfamiliar with the language of the generalized dielectric constant.

while the spectral function in such a model is evidently

$$g(\omega) = \int \delta(\omega_p(\mathbf{r}) - \omega)\, \varrho_0(\mathbf{r})\, d\mathbf{r}. \qquad (8.60)$$

The corresponding cross-section $\sigma(\omega)$ is calculated using the Thomas–Fermi atom density for $\varrho_0(r)$ in Fig. 8.7. The slow fall-off of the Thomas–Fermi density causes the cross-section to rise towards the constant value

$$\sigma_{\text{TF}}(0) = (3^3/2)^{\frac{1}{2}} \pi^3 e^2/mc = 46 \cdot 4 \text{ mb}. \qquad (8.61)$$

This then represents the Drude approximation to the photoabsorption cross-section, leading, as we remarked earlier, to a universal curve.

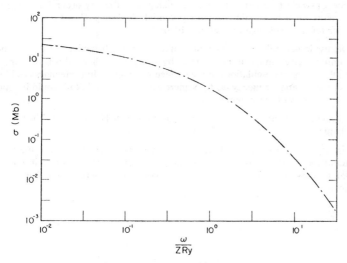

FIG. 8.7. Photo-absorption cross-section versus atomic number.

In concluding this chapter, it should be pointed out that collective modes in atoms have been successfully tackled by many-body techniques, in pioneering work by S. Lundqvist and his school in Gothenberg (see especially Wendin, 1972, and other references given there).[†]

[†] *Note added in proof.* The interested reader should also refer to the article by U. Fono and J. W. Cooper in *Rev. Mod. Phys.* **40**, 441, 1968.

By such techniques, Wendin has demonstrated that a broad collective resonance occurs in the collective behaviour of the $4d^{10}$ shell in Xe. The important conclusion here is that photoabsorption in this $4d^{10}$ shell is essentially a many-electron effect in which the photon energy is taken up by the shell as a whole. These calculations, which incorporate electron–electron correlations carefully into the ground-state (cf. Kelly, 1971) appear to allow the observations on photoabsorption to be understood at least semi-quantitatively.

Problems

1. For a two-electron ion work out the analogue of the low order $1/Z$ expansion coefficients for a harmonic oscillator potential replacing the Coulombic form $-Ze^2/r$ (cf. Byers-Brown and White, 1970).

2. Discuss in general terms how you would expect the behaviour of the plasmons in a uniform electron gas to be modified by a density gradient (this is relevant, for example, to plasma oscillations in the surface of a metal). In particular, would you expect the plasma frequency ω_p to be increased or reduced? And would it depend on the geometry of the system?

3. Calculate the probability of electron separation in the He atom from the Löwdin–Shull open shell wave function (5.32).

4. Use eqn. (4.5) for the kinetic energy of a uniform electron gas, and the corresponding eqn. (7.18) for the exchange energy, to derive the pressure p in the form (8.45). Discuss the effect on p of adding the correlation energy in the Wigner approximation.

CHAPTER 9

Relativistic effects in heavy atoms

WE have dealt so far entirely with the non-relativistic Schrödinger equation. However, it is not difficult to show that for the inner shell electrons of a heavy atom (e.g. the $1s$ electrons in Pb) the velocities are so high that relativistic effects become very important.

Though there remain some basic difficulties at the heart of a relativistic theory of a many-electron atom, successful approximate theories exist, which we shall consider quite briefly in this chapter.

In classical theory, the Hamiltonian H is equal to the energy E[†]

$$H = E \qquad (9.1)$$

and in quantum mechanics, since dynamical variables are associated with operators, we have instead of eqn. (9.1)

$$H \Psi = i\hbar \frac{\partial}{\partial t} \Psi \qquad (9.2)$$

where we have used the operator result expressing the fact that an Uncertainty Relation $\Delta E . \Delta t \sim \hbar$ exists between energy and time

$$E \rightarrow i\hbar \frac{\partial}{\partial t} . \qquad (9.3)$$

Then, with the Hamiltonian operator H obtained from the classical result for a single-particle moving in a potential energy $V(\mathbf{r})$

$$H = \frac{p^2}{2m} + V(\mathbf{r}) \qquad (9.4)$$

[†] See Appendix 5.1 for more details, and in particular for Hamilton's equations.

namely

$$H = -\frac{\hbar^2}{2m} \nabla^2 + V(\mathbf{r}), \tag{9.5}$$

space and time appear on a different footing. In particular, in eqn. (9.2) we have a first-order time derivative, whereas in the non-relativistic Hamiltonian (9.5), a second-order spatial operator ∇^2 appears.

Thus, the Schrödinger equation obtained by combining eqns. (9.2) and (9.5) cannot be invariant under a Lorentz transformation and is not compatible with the special theory of relativity.

9.1. Dirac wave equation

It is clear that in order to obtain a relativistic wave equation, one must start from the correct relativistic expression for the energy of a free particle, namely

$$(\text{Energy})^2 = c^2 p^2 + m^2 c^4 \tag{9.6}$$

where m is now the rest mass. For small momenta, this evidently yields

$$\text{Energy} = mc^2 \left[1 + \frac{1}{2} \frac{c^2 p^2}{m^2 c^4} + \cdots \right]$$

$$= mc^2 + \frac{p^2}{2m} 1 + \cdots \tag{9.7}$$

the first term being the energy associated with the rest mass according to the Einstein equivalence between mass and energy while the second term is simply the non-relativistic kinetic energy.

However, to form a wave equation by analogy with eqn. (9.1), we require not the square of the energy, as in eqn. (9.6), but the energy itself. Dirac (1928) therefore proposed that we write the quantity $p^2 + m^2 c^2$ appearing on the right-hand side of eqn. (9.6) as a perfect square, that is we put

$$p^2 + m^2 c^2 = p_x^2 + p_y^2 + p_z^2 + m^2 c^2 = (\alpha_x p_x + \alpha_y p_y + \alpha_z p_z + \alpha_m mc)^2. \tag{9.8}$$

It is then evident that the α's defined in eqn. (9.8) cannot be ordinary scalars, and we obtain the rules which the α's must satisfy by multiplying out the right-hand side of eqn. (9.8) and equating coefficients. It follows almost immediately that

$$\left.\begin{array}{ll} \alpha_x^2 = \alpha_y^2 = \alpha_z^2 = \alpha_m^2 = 1 & \\ \alpha_x\alpha_y = -\alpha_y\alpha_x, \quad \alpha_x\alpha_z = -\alpha_z\alpha_x, \quad \alpha_x\alpha_m = -\alpha_m\alpha_x \\ \alpha_y\alpha_z = -\alpha_z\alpha_y, \quad \alpha_y\alpha_m = -\alpha_m\alpha_y, \quad \alpha_z\alpha_m = -\alpha_m\alpha_z \end{array}\right\}. \qquad (9.9)$$

It is now clear that the α's must be operators, since they do not commute and provided we satisfy the relations (9.9) we can write the Hamiltonian H for a particle moving in a potential field V as

$$H = V - c\alpha_x p_x - c\alpha_y p_y - c\alpha_z p_z - \alpha_m mc^2. \qquad (9.10)$$

Here we have chosen the negative sign in taking the square root of eqn. (9.8). This is a pure convention and does not alter the physical content of eqns. (9.9) and (9.10).

Finally, we use the operator substitution (1.6) to obtain the Dirac Hamiltonian for an electron moving in a field of potential energy $V(\mathbf{r})$ as

$$H = V + ic\hbar\alpha_x\frac{\partial}{\partial x} + ic\hbar\alpha_y\frac{\partial}{\partial y} + ic\hbar\alpha_z\frac{\partial}{\partial z} - \alpha_m mc^2. \qquad (9.11)$$

While the theory strictly does not require that we take specific forms for the α's, it is usually much easier in practical problems to do so.

As in the Pauli spin theory (see, problem 9.3 and Schiff, 1955) it is convenient to choose the operators to be matrices. It can then be shown that 4×4 matrices are the smallest which can satisfy the relations (9.9). The set (not by any means unique) adopted by Dirac is

$$\alpha_x = \begin{pmatrix} 0 & 0 & 0 & 1 \\ 0 & 0 & 1 & 0 \\ 0 & 1 & 0 & 0 \\ 1 & 0 & 0 & 0 \end{pmatrix} \qquad \alpha_y = \begin{pmatrix} 0 & 0 & 0 & -i \\ 0 & 0 & i & 0 \\ 0 & -i & 0 & 0 \\ i & 0 & 0 & 0 \end{pmatrix}$$

$$\alpha_z = \begin{pmatrix} 0 & 0 & 1 & 0 \\ 0 & 0 & 0 & -1 \\ 1 & 0 & 0 & 0 \\ 0 & -1 & 0 & 0 \end{pmatrix} \qquad \alpha_m = \begin{pmatrix} 1 & 0 & 0 & 0 \\ 0 & 1 & 0 & 0 \\ 0 & 0 & -1 & 0 \\ 0 & 0 & 0 & -1 \end{pmatrix}. \qquad (9.12)$$

With the α's chosen as 4×4 matrices, the Dirac wave function Ψ in the wave eqn. (9.2) must be a column vector having four components, namely

$$\Psi = \begin{pmatrix} \psi_1 \\ \psi_2 \\ \psi_3 \\ \psi_4 \end{pmatrix}. \qquad (9.13)$$

Then we obtain from eqns. (9.2) and (9.11)−(9.13) four ordinary coupled differential equations to solve for the components $\psi_1 - \psi_4$ of the column vector (9.13). These equations have the explicit form

$$\left. \begin{aligned} \left(\frac{i}{\hbar}\right)\left[\frac{E-V}{c}+mc\right]\psi_1+\left[\frac{\partial}{\partial x}-i\frac{\partial}{\partial y}\right]\psi_4+\frac{\partial\psi_3}{\partial z} = 0 \\[4pt] \left(\frac{i}{\hbar}\right)\left[\frac{E-V}{c}+mc\right]\psi_2+\left[\frac{\partial}{\partial x}+i\frac{\partial}{\partial y}\right]\psi_3-\frac{\partial\psi_4}{\partial z} = 0 \\[4pt] \left(\frac{i}{\hbar}\right)\left[\frac{E-V}{c}-mc\right]\psi_3+\left[\frac{\partial}{\partial x}-i\frac{\partial}{\partial y}\right]\psi_2+\frac{\partial\psi_1}{\partial z} = 0 \\[4pt] \left(\frac{i}{\hbar}\right)\left[\frac{E-V}{c}-mc\right]\psi_4+\left[\frac{\partial}{\partial x}+i\frac{\partial}{\partial y}\right]\psi_1-\frac{\partial\psi_2}{\partial z} = 0 \end{aligned} \right\}. \qquad (9.14)$$

9.2. Central field solutions of Dirac equation

Our interest here is in the solution of eqns. (9.14) when the potential energy $V(\mathbf{r})$ is spherically symmetric, that is $V(\mathbf{r}) = V(|\mathbf{r}|)$. As with the Schrödinger equation, discussed for central fields in Chapter 2, spherical polar coordinates (r, θ, ϕ) are the most natural ones to use. Thus, in eqns. (9.14), we use the relations

$$\frac{\partial}{\partial x}+i\frac{\partial}{\partial y} = e^{i\phi}\left[\sin\theta\,\frac{\partial}{\partial r}+\frac{\cos\theta}{r}\,\frac{\partial}{\partial\theta}+\frac{i}{r\sin\theta}\,\frac{\partial}{\partial\phi}\right] \quad (9.15)$$

$$\frac{\partial}{\partial z} = \cos\theta\,\frac{\partial}{\partial r}-\frac{\sin\theta}{r}\,\frac{\partial}{\partial\theta}, \qquad (9.16)$$

while $(\partial/\partial x)-i(\partial/\partial y)$ is obtained from eqn. (9.15), by replacing i by $-i$ on the right-hand side.

The resulting equations can then be separated, as in the non-relativistic case. However, since electron spin and magnetic moment is now included in the Dirac equation itself, unlike the Schrödinger case, where the Pauli spin theory has to be grafted on, the total angular momentum **J** plays the role for the Dirac equation that the orbital angular momentum **L** played for the Schrödinger equation.

Hence we seek a wave function Ψ satisfying the equations

$$J^2\Psi = j(j+1)\hbar^2\Psi, \qquad j = \tfrac{1}{2}, \tfrac{3}{2} \ldots, \qquad (9.17)$$

j being half-integral as spin is now included, and

$$J_z\Psi = m\hbar\Psi, \qquad m = \pm\tfrac{1}{2}, \pm\tfrac{3}{2}. \qquad (9.18)$$

with $j \geqslant |m|$. The solutions Ψ for a given m and j then take the form

$$
\left.
\begin{aligned}
\psi_1 &= f(r)\left(\frac{j+1-m}{2j+2}\right)^{\frac{1}{2}} Y_{j+\frac{1}{2},\,m-\frac{1}{2}}(\theta,\phi) + g(r)\left(\frac{j+m}{2j}\right)^{\frac{1}{2}} Y_{j-\frac{1}{2},\,m-\frac{1}{2}}(\theta,\phi) \\[2ex]
\psi_2 &= -f(r)\left(\frac{j+1+m}{2j+2}\right)^{\frac{1}{2}} Y_{j+\frac{1}{2},\,m+\frac{1}{2}}(\theta,\phi) + g(r)\left(\frac{j-m}{2j}\right)^{\frac{1}{2}} Y_{j-\frac{1}{2},\,m+\frac{1}{2}}(\theta,\phi) \\[2ex]
\psi_3 &= F(r)\left(\frac{j+1-m}{2j+2}\right)^{\frac{1}{2}} Y_{j+\frac{1}{2},\,m-\frac{1}{2}}(\theta,\phi) + G(r)\left(\frac{j+m}{2j}\right)^{\frac{1}{2}} Y_{j-\frac{1}{2},\,m-\frac{1}{2}}(\theta,\phi) \\[2ex]
\psi_4 &= -F(r)\left(\frac{j+1+m}{2j+2}\right)^{\frac{1}{2}} Y_{j+\frac{1}{2},\,m+\frac{1}{2}}(\theta,\phi) + G(r)\left(\frac{j-m}{2j}\right)^{\frac{1}{2}} Y_{j-\frac{1}{2},\,m+\frac{1}{2}}(\theta,\phi)
\end{aligned}
\right\}
$$

$$(9.19)$$

where, as discussed earlier, $Y_{lm}(\theta,\phi)$ are the spherical harmonics, which are related explicitly to the associated Legendre functions $P_l^m(\cos\theta)$ by (see also eqns. 1.34 and 1.36)

$$Y_{lm}(\theta,\phi) = (-1)^m \left\{ \frac{(2l+1)\,(l-m)!}{4\pi(l+m)!} \right\}^{\frac{1}{2}} P_l^m(\cos\theta)e^{im\phi} \qquad (9.20)$$

Substituting these results into eqns. (9.14) written in spherical polar coordinates, it is readily shown that the four functions f, g, F and G

satisfy the radial equations

$$
\left.
\begin{aligned}
\frac{df}{dr} + \left(j + \frac{3}{2}\right)\frac{f}{r} + \frac{i}{\hbar c}[E - V(r) - mc^2]G &= 0 \\[2mm]
\frac{dG}{dr} - \left(j - \frac{1}{2}\right)\frac{G}{r} + \frac{i}{\hbar c}[E - V(r) + mc^2]f &= 0 \\[2mm]
\frac{dF}{dr} + \left(j + \frac{3}{2}\right)\frac{F}{r} + \frac{i}{\hbar c}[E - V(r) + mc^2]g &= 0 \\[2mm]
\frac{dg}{dr} - \left(j - \frac{1}{2}\right)\frac{g}{r} + \frac{i}{\hbar c}[E - V(r) - mc^2]F &= 0
\end{aligned}
\right\}
\tag{9.21}
$$

Thus, again we have separated the wave equation, and the problem in self-consistent field theory is to solve the above equations for a chosen $V(r)$ and continue iteratively until the wave functions calculated from $V(r)$ reproduce the same $V(r)$, in the Hartree sense.

In Appendix 9.1 it is shown that, for the Coulomb field, with $V(r) = -Ze^2/r$, eqns. (9.21) can be solved analytically, and the famous Sommerfeld formula for the energy level spectrum results.

It is worth emphasizing that eqns. (9.21) now fall into pairs, the first pair being simultaneous equations to be solved for f and G (see Appendix 9.1 for explicit solutions for the Coulomb field).

By now, a solution of the Hartree problem using the relativistic eqns. (9.21) has been carried out for a number of atoms and ions (one example is the calculation by Williams (1940) on Cu^+).

If we wished to transcend single-particle theories, then we should have to use the appropriate many-particle wave equation, with proper relativistic interactions between electrons. The precise expressions are not known, but for heavy atoms, for which relativistic effects need to be considered, the main effects appear to arise firstly from the relativistic variation of mass with velocity and secondly from the spin–orbit interaction.

Fortunately, these two effects may be included satisfactorily using Dirac wave functions and the simple electrostatic form e^2/r_{ij} for the electron–electron interaction,[†] the total interaction being a sum over pairs ij. This approximation neglects the fact that electromagnetic disturbances propagate with finite velocity c: that is retardation effects are neglected.

A full relativistic many-electron theory is therefore still lacking, but in practice we can obtain results of high accuracy in heavy atoms using the Dirac equation.

9.3. Relativistic Thomas–Fermi theory

For heavy atoms, it is tempting to try to extend the Thomas–Fermi theory of Chapter 4 to include relativistic effects. The earliest attempt to do so appears to be that of Vallarta and Rosen (1932): these workers included the relativistic variation of mass with velocity.

Such Thomas–Fermi theories have not been notably successful to date and we shall therefore restrict ourselves to a very brief summary here.

Following the procedure outlined in introducing the Dirac equation we write

$$c^2 p^2 + m^2 c^4 = [E + mc^2 - V]^2 \qquad (9.22)$$

where $E + mc^2$ is the total energy. As in the non-relativistic Thomas–Fermi atom, we argue that the maximum energy E of a bound electron is zero and hence we find for the maximum momentum p_f the result

$$\frac{p_f^2(\mathbf{r})}{2m} + \frac{V^2}{2mc^2} + V = 0. \qquad (9.23)$$

It is clear that as $c \to \infty$ the previous relation between p_f and potential energy V is regained. But, as before

$$\varrho(r) = \frac{8\pi}{3h^3} p_f^3 \equiv \frac{8\pi}{3h^3} \left\{ \frac{V^2}{c^2} - 2mV \right\}^{\frac{3}{2}}. \qquad (9.24)$$

[†] See, for example, the discussion of Brown (1952), for a more precise form of the interaction, or the later work of Layzer and Bahcall (1962).

Hence from Poisson's equation we find

$$\nabla^2 V = \frac{32\pi^2}{3h^3}\left\{-2mV+\frac{V^2}{c^2}\right\}^{\frac{3}{2}}.\qquad(9.25)$$

There are more serious troubles with divergence of the charge density near the nucleus than with the non-relativistic equation. However, Vallarta and Rosen (1932) assumed that the term (V^2/c^2) could be treated as a perturbation, and solved for the correction to the self consistent field of Thomas and Fermi for mercury $(Z = 80)$. The corrections were found to be quite small, except near the nucleus, where we have already remarked on the troubles which beset the theory.

It is clear that the correct generalization of the Thomas–Fermi theory for electrons in an atom must be based on the Dirac wave equation. Rudkjøbing (1952) has derived the density of states resulting from the Dirac equation for the case of a spherically symmetric potential and Gilvarry (1954) has utilized this result to derive a relativistic Thomas–Fermi equation for an atom based on the Dirac equation. We shall briefly summarize the main results, without proof, below, and comment on the relation to the equation of Vallarta and Rosen.

The number of states $\varrho(r, E)$ per unit volume and per unit energy range for an electron of total energy E at a point r in a spherically symmetric atom with potential energy $V(r)$ may according to Rudkjøbing be written as

$$\varrho(r,E) = \frac{8\pi}{h^3c^3}\left[(E-V)^2-m^2c^4-\left(r\frac{dV}{dr}\right)^2\right]^{\frac{1}{2}}[E-V]\qquad(9.26)$$

The energy term $(r\,dV/dr)^2$ in the density $\varrho(r,E)$ is interpreted by Rudkjøbing as due to spin–orbit interaction, since it arises from the terms in the Dirac equation which yield this effect quantum mechanically.

The desired relativistic Thomas–Fermi equation may then be written

$$\frac{1}{r}\frac{d^2}{dr^2}(rV) = \frac{32\pi^2}{3}\frac{(2m)^{\frac{3}{2}}}{h^3}$$

$$\times\left[(E_f-mc^2)-V+\left\{(E_f-mc^2-V)^2-\left(r\frac{dV}{dr}\right)^2\right\}\Big/2mc^2\right]^{\frac{3}{2}}.\qquad(9.27)$$

If we put $E_f - mc^2$ equal to zero, and omit the spin–orbit term $(r\,dV/dr)^2$, this equation reduces to eqn. (9.25). Again the equation can be solved by perturbation theory, based on the usual non-relativistic Thomas–Fermi solution as the zeroth order approximation.

(a) *Variation of atomic binding energies with Z*

It might be of interest to apply the above equation to estimate the correction to the binding energy of the Thomas–Fermi atom due to relativistic effects. To our knowledge, this has not been done, and therefore we conclude this brief discussion by referring to some approximate numerical estimates of these relativistic corrections (Scott, 1952).

It appears that relativistic effects on the binding energies of atoms have no serious consequences for $Z \leqslant 30$.

Scott points out that the relativistic correction to the energy required to strip off the last two electrons is accurately known, and that the correction for stripping off all the others is only of the same order of magnitude. Scott's numerical estimates lead him to the approximate expression

$$|E|_{\text{relativistic}} - |E|_{\text{non-relativistic}} = 4 \times 10^{-6} Z^{\frac{9}{2}} \frac{e^2}{a_0}. \qquad (9.28)$$

This expression yields about 15 au at Cu to $\sim$ 2000 au at U.[†]

But our concluding remarks must take us back to the $1/Z$ expansion, and to a much more basic approach to calculating relativistic corrections. As is clear from the discussion of the $1/Z$ expansion, we can express atomic energy levels in the form, for the non-relativistic case,

$$E_{\text{non-relativistic}} = Z^2 \sum_{n=0}^{\infty} \varepsilon_{n0} Z^{-n} \qquad (9.29)$$

which, as we emphasized, has the advantage over the numerical Hartree method of determining a given energy level simultaneously for all members of an isoelectronic sequence. Layzer and Bahcall (1962)

[†] Scott's estimate of the error in the total binding energy resulting from the neglect of relativistic effects is $(Z/30)^2$ per cent.

extended the $1/Z$ expansion to include relativistic corrections. The relativistic energy has now a double power series expansion in Z^{-1} and in $\alpha^2 Z^2$, namely

$$E = Z^2 \sum_{n=0}^{\infty} \sum_{m=0}^{\infty} \varepsilon_{nm} Z^{-n}(\alpha^2 Z^2)^m, \qquad (9.30)$$

$\alpha = e^2/\hbar c$ the fine structure constant. The coefficients ε_{nm} are functions of $\alpha^2 Z^3$. This approach has been reviewed by Doyle (1969) and we must refer the reader to his article for details. Excellent agreement between theory and experiment has resulted from this method

Problems

1. Develop the Sommerfeld fine-structure formula (A9.1.16) for an atom of charge Z, and discuss the corrections to the non-relativistic theory, by means of a α/Z expansion.

2. By considering the energy terms arising from the motion of an electron obeying the Dirac equation (9.11) in a uniform magnetic field $\mathcal{H}$, identify a term of the form $\boldsymbol{\mu} \cdot \mathcal{H}$ and hence obtain the magnetic moment associated with electron spin.

(*Hints:* (i) It will be useful to obtain the energy in a suitable non-relativistic limit in identifying $\boldsymbol{\mu}$. (ii) The necessary commutation relations for angular momentum are given in problem 5 of Chapter 1, see also problem 3 below.)

3. The Pauli spin matrices σ_x, σ_y and σ_z defined by (see, for example, Mott and Sneddon, 1948)

$$\sigma_x = \begin{pmatrix} 0 & i \\ i & 0 \end{pmatrix}, \qquad \sigma_y = \begin{pmatrix} 0 & -i \\ i & 0 \end{pmatrix}, \qquad \sigma_z = \begin{pmatrix} i & 0 \\ 0 & -i \end{pmatrix}$$

are related to the spin angular momentum $\mathbf{s} = (s_x, s_y, s_z)$ by $\mathbf{s} = \frac{1}{2}\hbar\boldsymbol{\sigma}$, with $= \boldsymbol{\sigma} = (\sigma_x, \sigma_y, \sigma_z)$.

(i) Show that the commutation rules for these spin operators are (compared problem 1.5 for orbital angular momentum $\mathbf{L}$)

$$s_x s_y - s_y s_x = i\hbar s_z, \quad s_y s_z - s_z s_y = i\hbar s_x, \quad s_z s_x - s_x s_z = i\hbar s_y.$$

(ii) show that $\sigma_x^2 = \sigma_y^2 = \sigma_z^2 = 1$ and discuss their eigenvalues.

(iii) If $\mathbf{J}$ is the total angular momentum $\mathbf{L} + \mathbf{s}$ of an electron, show that the Pauli operators associated with J_z and J^2 are

$$J_z = \begin{pmatrix} L_z + \frac{1}{2}\hbar & 0 \\ 0 & L_z - \frac{1}{2}\hbar \end{pmatrix}$$

and

$$J^2 = \begin{pmatrix} L^2 + \hbar L_z + \frac{3}{4}\hbar^2 & \hbar(L_x - iL_y) \\ \hbar(L_x + iL_y) & L^2 - \hbar L_z + \frac{3}{4}\hbar^2 \end{pmatrix}.$$

(iv) For $j = m = \frac{1}{2}$, use (iii) above to show that (9.19) represent eigenfunctions of J^2 and J_z.

APPENDIX 1.1

Orthogonality of solutions of Schrödinger equation

WE prove the orthogonality of solutions of the Schrödinger equation for one dimension. In the three-dimensional case, solutions of the wave equation corresponding to the same degenerate level need not be orthogonal.

The wave equation for the nth level takes the form

$$\frac{d^2\psi_n}{dx^2} + \frac{2m}{\hbar^2}[E_n - V(x)]\psi_n = 0. \qquad (A1.1.1)$$

Since $V(x)$ is real, we can immediately write down the following identities from eqn. (A1.1.1):

$$\psi_m^* \frac{d^2\psi_n}{dx^2} + \frac{2m}{\hbar^2}\psi_m^*[E_n - V(x)]\psi_n = 0 \qquad (A1.1.2)$$

and

$$\psi_n \frac{d^2\psi_m^*}{dx^2} + \frac{2m}{\hbar^2}\psi_n[E_m - V(x)]\psi_m^* = 0. \qquad (A1.1.3)$$

By subtraction of eqn. (A1.1.3) from eqn. (A1.1.2), and integration over all space we eliminate the potential energy $V(x)$ to find

$$\int \left[\psi_m^* \frac{d^2\psi_n}{dx^2} - \psi_n \frac{d^2\psi_m^*}{dx^2} \right] dx$$

$$+ \frac{2m}{\hbar^2}(E_n - E_m) \int \psi_m^* \psi_n \, dx = 0. \qquad (A1.1.4)$$

We now use the identity

$$\frac{d}{dx}\left[\psi_m^* \frac{d\psi_n}{dx} - \psi_n \frac{d\psi_m^*}{dx}\right]$$
$$= \psi_m^* \frac{d^2\psi_n}{dx^2} - \psi_n \frac{d^2\psi_m^*}{dx^2}. \qquad (A1.1.5)$$

We can then integrate over x and, provided

$$\psi_m^* \frac{d\psi_n}{dx} - \psi_n \frac{d\psi_m^*}{dx}$$

vanishes at the limits, as is usually the case, we can write

$$(E_n - E_m)\int \psi_m^* \psi_n \, dx = 0. \qquad (A1.1.6)$$

Thus, for different energy levels, that is $E_n \neq E_m$, we have

$$\int \psi_m^* \psi_n \, dx = 0 \qquad (A1.1.7)$$

which is the orthogonality property we wished to prove.

APPENDIX 2.1

Radial wave functions for hydrogen atom

STARTING from a total wave function

$$\psi(r, \theta, \phi) = R(r) Y_{lm}(\theta, \phi) \qquad (A2.1.1)$$

we showed that, with $P = rR$ (see eqn. 1.13, with $V(r) = -e^2/r$ and $\mu = l(l+1)$)

$$\frac{d^2P}{dr^2} + \frac{2m}{\hbar^2}\left[E + \frac{e^2}{r} - \frac{\hbar^2}{2m}\frac{l(l+1)}{r^2}\right]P = 0. \qquad (A2.1.2)$$

Put $E = -|E|$ since E is negative for bound states. Also write

$$\varrho = 2\left[\sqrt{\left(\frac{2m}{\hbar^2}|E|\right)}\right]r; \qquad \lambda = \frac{me^2}{\hbar^2}\bigg/\sqrt{\frac{2m}{\hbar^2|E|}}\right) \qquad (A2.1.3)$$

and then from (A2.1.2) and (A.2.1.3) we find

$$\frac{d^2P}{d\varrho^2} + \left[-\frac{1}{4} + \frac{\lambda}{\varrho} - \frac{l(l+1)}{\varrho^2}\right]P = 0. \qquad (A2.1.4)$$

Now use the asymptotic solution $p \sim e^{-\varrho/2}$ for large ϱ, to write for all ϱ

$$P = \varrho^{l+1}e^{-\varrho/2}f(\varrho). \qquad (A2.1.5)$$

We then obtain from (A2.1.4) after a simple calculation (with $f' = df/d\varrho$ etc.)

$$\varrho f'' + 2f'[l+1 - \tfrac{1}{2}\varrho] + [\lambda - (l+1)]f = 0. \qquad (A2.1.6)$$

Now we expand $f(\varrho)$ in the series form

$$f(\varrho) = a_\tau\varrho^\tau + a_{\tau+1}\varrho^{\tau+1} + \ldots \qquad (A2.1.7)$$

with $a_\tau \neq 0$. Then we have

$$f'' = \tau(\tau-1)a_\tau \varrho^{\tau-2}+(\tau+1)(\tau)a_{\tau+1}\varrho^{\tau-1}+ \ldots \quad (A2.1.8)$$

Therefore the coefficient of $\varrho^{\tau-1}$ in (A2.1.6) is

$$\tau(\tau-1)a_\tau+2\tau a_\tau(l+1) = 0. \quad (A2.1.9)$$

Hence it follows that if $a_\tau \neq 0$,

$$\tau = -(2l+1) \text{ or } \tau = 0, \quad (A2.1.10)$$

We must take $\tau = 0$ for the well-behaved solutions as $\varrho \to 0$. Now, with $\tau = 0$,

$$\left.\begin{aligned}
f(\varrho) &= a_0+a_1\varrho+ a_2\varrho^2+\ldots+a_\nu\varrho^\nu &&+a_{\nu+1}\varrho^{\nu+1} &&+ \ldots \\
f'(\varrho) &= a_1 +2a_2\varrho +\ldots+\nu a_\nu\varrho^{\nu-1} &&+(\nu+1)a_{\nu+1}\varrho^\nu &&+ \ldots \\
f''(\varrho) &= 2a_2 +\ldots+\nu(\nu-1)a_\nu\varrho^{\nu-2} &&+(\nu+1)\nu a_{\nu+1}\varrho^{\nu-1} &&+ \ldots
\end{aligned}\right\}$$
$$(A2.1.11)$$

At this stage we pick out the coefficient of ϱ^ν in (A2.1.6). This is

$$(\nu+1)\nu a_{\nu+1}+2(l+1)(\nu+1)a_{\nu+1}-\nu a_\nu$$
$$+a_\nu[\lambda-(l+1)] = 0. \quad (A2.1.12)$$

Thus we find

$$(\nu+1)(\nu+2l+2)a_{\nu+1} = -a_\nu(\lambda-\nu-l-1) \quad (A2.1.13)$$

or

$$\frac{a_{\nu+1}}{a_\nu} = -\frac{(\lambda-\nu-l-1)}{(\nu+1)(\nu+2l+2)}. \quad (A2.1.14)$$

The series for f terminates if $\lambda = \nu+l+1$, $\nu = 0, 1, 2$, etc., that is if and only if λ is an integer greater than zero, say $\lambda = n$; $n = 1, 2, 3$, etc. n is, of course, the principal quantum number. The necessity for the series for f to terminate is seen by noting from eqn. (A2.1.14) that

$$\frac{a_{\nu+1}}{a_\nu} \sim \frac{1}{\nu} \quad \text{as} \quad \nu \to \infty \quad (A2.1.15)$$

and if we compare this with the coefficients in the series

$$e^{\varrho} = 1 + \varrho + \frac{\varrho^2}{2!} + \cdots \qquad (A2.1.16)$$

we find these behave as in eqn. (A2.1.15). Thus $f(\varrho)$ diverges as e^{ϱ} as $\varrho \to \infty$, and hence the wave function diverges unless the series for $f(\varrho)$ terminates. This of course determines the allowed values of the energy (related to λ) in terms of the principal quantum number n, as in eqn. (1.2).

The polynomial $f(\varrho)$, starting with a term in ϱ^0 and terminating at ϱ^{ν}: that is ϱ^{n-l-1} turns out to be the polynomial

$$\frac{d^{2l+1}}{d\varrho^{2l+1}} \left[e^{\varrho} \frac{d^{n+l}}{d\varrho^{n+l}} (\varrho^{n+l} e^{-\varrho}) \right]. \qquad (A2.1.17)$$

Hence $P = rR$ is given by

$$P(nl) = \varrho^{l+1} e^{-\varrho/2} L_{n+l}^{2l+1}(\varrho) \qquad (A2.1.18)$$

where $L_{n+l}^{2l+1}(\varrho)$ is the associated Laguerre function and ϱ is simply $2r/na_0$, when we make use of eqns. (A2.1.3) and (1.2).

APPENDIX 2.2

Variation method

SUPPOSE that we have a system described by a Hamiltonian H. The wave equation may be written

$$H\psi = E\psi \tag{A2.2.1}$$

and in particular, if ψ_0 is the wave function and E_0 the energy of the lowest state, then

$$H\psi_0 = E_0\psi_0. \tag{A2.2.2}$$

In quantum mechanics the energy of the ground state is often of particular interest: and the variation method enables us to approximate to it.

Consider any function ϕ which is well behaved and we will assume also normalized. Now let us form the quantity $\mathcal{E}$ defined by

$$\mathcal{E} = \int \phi^* H \phi \, d\tau. \tag{A2.2.3}$$

We show below that

$$\mathcal{E} \geqslant E_0, \tag{A2.2.4}$$

which is the basis of the variation method.

PROOF

If ϕ is not equal to ψ_0 we can expand ϕ in terms of the complete set of normalized, orthogonal functions, $\psi_0, \psi_1 \ldots \psi_n \ldots$, which are the solutions of (A2.2.1), obtaining for a normalized ϕ,

$$\phi = \sum_n a_n \psi_n$$

with
$$\sum_n a_n^* a_n = 1. \tag{A2.2.5}$$

Therefore from eqns. (A2.2.3) and (A2.2.5)

$$\mathcal{E} = \int \sum_{n'} a_{n'}^* \psi_{n'}^* H \sum_n a_n \psi_n \, d\tau \tag{A2.2.6}$$

But from eqn. (A2.2.1) $H\psi_n = E\psi_n \tag{A2.2.7}$

Thus it follows that

$$\mathcal{E} = \int \sum_{n'} a_{n'}^* \psi_{n'}^* \sum_n E_n a_n \psi_n \, d\tau. \tag{A2.2.8}$$

But again $\int \psi_{n'}^* \psi_n \, d\tau = 0$ if $n' \neq n$,   and $= 1$ if $n' = n$,

and therefore

$$\mathcal{E} = \sum_n a_n^* a_n E_n. \tag{A2.2.9}$$

Using eqn. (A2.2.5) this can be written

$$\mathcal{E} - E_0 = \sum_n a_n^* a_n (E_n - E_0). \tag{A2.2.10.}$$

But E_0 is the lowest energy level by definition, and hence

$$E_n - E_0 \geqslant 0. \tag{A2.2.11}$$

Clearly $a_n^* a_n \geqslant 0, \tag{A2.2.12}$

and hence, from eqn. (A2.2.10) we obtain the desired result (A2.2.4).

APPENDIX 2.3

Hartree equations derived from a variational principle

WE take the expectation value of the Hamiltonian with respect to the wave function $\Pi_i \phi_i^\dagger$ and determine the ϕ_i's by finding the conditions that the energy be stationary with respect to variations in the ϕ_i's subject only to the condition that the single-particle functions remain normalized.

Since there are N such wave functions, we need N Lagrange multipliers and hence we consider variations of the quantity

$$E + \sum_{i=1}^{N} \lambda_i \int \phi_i^* \phi_i \, d\mathbf{r}_i \qquad (A2.3.1)$$

where

$$\begin{aligned}
E = &\sum_{i=1}^{N} \int \phi_i^* H_i \phi_i \, d\mathbf{r}_i \\
&+ \frac{1}{2} e^2 \sum_{ij(i \neq j)} \int \int \phi_i^*(\mathbf{r}_i) \, \phi_j^*(\mathbf{r}_j) \frac{1}{r_{ij}} \, \phi_i(\mathbf{r}_i) \, \phi_j(\mathbf{r}_j) \, d\mathbf{r}_i \, d\mathbf{r}_j.
\end{aligned} \qquad (A2.3.2)$$

We now make an arbitrary variation in one of the ϕ_i, and strictly in ϕ_i^* but, as with the elementary example given above, the coefficient of $\delta\phi_i^*$ will be the complex conjugate of that of $\delta\phi_i$ and we do not need to consider both variations.

Hence we have

---

† $\Pi_i \phi_i$ is a notation for a total wave function which is a simple product of the one-electron wave functions ϕ_i

$$\int \delta\phi_i^* \left[H_i\phi_i(\mathbf{r}_i) + e^2 \sum_{j \neq i} \int \phi_j^*(\mathbf{r}_j) \frac{1}{r_{ij}} \phi_j(\mathbf{r}_j) \, d\mathbf{r}_j \phi + \lambda_i\phi_i(\mathbf{r}_i) \right] d\mathbf{r}_i = 0$$

(A2.3.3)

Again, since $\delta\phi_i^*$ is an arbitrary variation we must have

$$\left[H_i + e^2 \sum_{j \neq i} \int \phi_j^*(\mathbf{r}_j) \frac{1}{r_{ij}} \phi_j(\mathbf{r}_j) \, d\mathbf{r}_j + \lambda_i \right] \phi_i(\mathbf{r}_i) = 0. \quad (A2.3.4)$$

We identify $\lambda_i = -\varepsilon_i$, the single-particle energies, and we see that the equations which determine the best single-particle functions are the Hartree equations already referred to, namely

$$\left[-\frac{\hbar^2}{2m} \nabla_i^2 - \frac{Ze^2}{r_i} + e^2 \sum_{j \neq i} \int \frac{|\phi_j(\mathbf{r}_j)|^2}{r_{ij}} \, d\mathbf{r}_j \right] \phi_i(\mathbf{r}_i) = \varepsilon_i\phi_i(\mathbf{r}_i). \quad (A2.3.5)$$

Thus, the basic eqns. (2.53) of the self-consistent field theory are regained: their physical content being exactly what Hartree had anticipated in his original paper (reprint 1 of this volume).

APPENDIX 3.1

Solutions of the dimensionless Thomas–Fermi equation

See, for example, Miranda, 1934; Gombás, 1949)

$$\frac{d^2\phi}{dx^2} = \frac{\phi^{\frac{3}{2}}}{x^{\frac{1}{2}}} \quad \text{with} \quad \phi(0) = 1$$

Initial slope $(-\phi'(0))$	1·60 $x_0 = 2·795$	1·589 $x_0 = 6·423$	1·5882 $x_0 = 9·7558$	1·58805 (neutral atom solution)
x	$\phi(x)$			
0·00	1·0000	1·0000	1·0000	1·000
0·10	0·8803	0·8816	0·8817	0·882
0·20	0·7892	0·7929	0·7930	0·793
0·30	0·7142	0·7203	0·7206	0·721
0·40	0·6502	0·6591	0·6595	0·660
0·50	0·5947	0·6064	0·6069	0·607
0·60	0·5456	0·5605	0·5611	0·561
0·70	0·5018	0·5199	0·5207	0·521
0·80	0·4623	0·4839	0·4848	0·485
0·90	0·4263	0·4517	0·4527	0·453
1·00	0·3932	0·4227	0·4238	0·425
1·10	0·3626	0·3964	0·3977	
1·20	0·3341	0·3725	0·3740	0·375
1·30	0·3073	0·3507	0·3524	
1·40	0·2821	0·3307	0·3326	0·333
1·50	0·2581	0·3123	0·3144	
1·60	0·2352	0·2954	0·2977	0·298
1·70	0·2131	0·2797	0·2823	
1·80	0·1919	0·2651	0·2681	0·268
1·90	0·1712	0·2516	0·2548	
2·00	0·1511	0·2390	0·2424	0·242

Initial slope $(-\phi'(0))$	1·60 $x_0 = 2\cdot795$	1·589 $x_0 = 6\cdot423$	1·5882 $x_0 = 9\cdot7558$	1·58805 (neutral atom solution)
2·20	0·1120	0·2162	0·2203	0·220
2·40	0·0740	0·1961	0·2009	0·201
2·60	0·0364	0·1782	0·1839	0·185
2·80		0·1623	0·1688	0·171
3·00		0·1479	0·1554	0·158
3·20		0·1348	0·1434	0·146
3·40		0·1228	0·1326	0·135
3·60		0·1117	0·1229	0·125
3·80		0·1015	0·1141	0·116
4·00		0·0919	0·1061	0·108
4·50		0·0701	0·0888	0·0918
5·00		0·0506	0·0748	0·0787
5·50		0·0323	0·0630	0·0679
6·00		0·0148	0·0530	0·0592
6·50			0·0441	0·0521
7·00			0·0362	0·0461
7·50			0·0290	0·0409
8·00			0·0222	0·0365
8·50			0·0157	0·0327
9·00			0·0094	0·0295
9·50			0·0032	0·0268
10·0				0·0244
10·5				0·0223
11·0				0·0204
11·5				0·0187
12·0				0·0172
12·5				0·0159
13·0				0·0147
13·5				0·0136
14·0				0·0126
14·5				0·0117
15·0				0·0109

Hamiltonian for charged particle in an electromagnetic field

CONSIDER an electron, charge $-e$, rest mass m, in an electromagnetic field, with fields $\mathcal{E}$ and $\mathcal{H}$.

From electromagnetic theory the force $\mathbf{F}$ on the charged particle is

$$\mathbf{F} = -e\mathcal{E} - \frac{e\mathbf{v} \times \mathcal{H}}{c} \qquad (\text{A}5.1.1)$$

In (A5.1.1) the velocity of the electron is $\mathbf{v}$. The equations of motion in non-relativistic theory[+] are $(d/dt)(m\mathbf{v}) = \mathbf{F}$ and taking account of the variation of mass with velocity, namely

$$m \bigg/ \sqrt{\left(1 - \frac{v^2}{c^2}\right)}.$$

The relativistic equations of motion are

$$\frac{d}{dt}\left(\frac{m\mathbf{v}}{\sqrt{\left(1 - \frac{v^2}{c^2}\right)}}\right) = -e\mathcal{E} - \frac{e\mathbf{v} \times \mathcal{H}}{c}. \qquad (\text{A}5.1.2)$$

The Hamiltonian by definition is the energy written in terms of coordinates and conjugate momenta.

If we know the Hamiltonian H, in terms of x and conjugate momentum p_x, then we have Hamilton's equations (cf. Pauling and Wilson, 1935)

[+] In this Appendix, Chapter 9 and Appendix 9.1, relativistic equations are employed, in contrast to the remainder of the book.

$$\frac{\partial H}{\partial p_x} = \dot{x} \quad \text{and} \quad \frac{\partial H}{\partial x} = -\dot{p}_x \qquad \text{(A5.1.3)}$$

and any Hamiltonian we set up must satisfy (A5.1.3).

The Hamiltonian we are seeking is in fact

$$H = -e\phi + c\sqrt{\left[\left(\mathbf{p} + \frac{e}{c}\mathbf{A}\right)^2 + m^2c^2\right]}, \qquad \text{(A5.1.4)}$$

with ϕ and $\mathbf{A}$ the scalar and vector potentials respectively, and we will now verify this. What we shall do is to show that this Hamiltonian, together with eqns. (A5.1.3), leads to the equations of motion (A5.1.2).

Proof that (A5.1.4) is correct Hamiltonian.

From eqn. (A5.1.3)

$$\dot{x} = \frac{dx}{dt} = v_x = \frac{\partial H}{\partial p_x} \qquad \text{(A5.1.5)}$$

and using H as in (A5.1.4) we find

$$v_x = c\left(p_x + \frac{e}{c}A_x\right)\Big/\sqrt{\left[\left(\mathbf{p} + \frac{e}{c}\mathbf{A}\right)^2 + m^2c^2\right]}. \qquad \text{(A5.1.6)}$$

Also from eqns. (A5.1.3) and (A5.1.4)

$$\dot{p}_x = \frac{dp_x}{dt} = -\frac{\partial H}{\partial x} = e\frac{\partial \phi}{\partial x} - \frac{e}{c}\left(v_x\frac{\partial A_x}{\partial x} + v_y\frac{\partial A_y}{\partial x} + v_z\frac{\partial A_z}{\partial x}\right) \qquad \text{(A5.1.7)}$$

where use has been made of eqn. (A5.1.6) for v_x. Now we can write

$$\frac{dA_x}{dt} = \frac{\partial A_x}{\partial t} + \frac{\partial A_x}{\partial x}v_x + \frac{\partial A_x}{\partial y}v_y + \frac{\partial A_x}{\partial z}v_z \qquad \text{(A5.1.8)}$$

and if we multiply this equation by e/c and add to eqn. (A5.1.7) we find

$$\frac{d}{dt}\left(p_x + \frac{e}{c}A_x\right) = e\frac{\partial\phi}{\partial x} + \frac{e}{c}\frac{\partial A_x}{\partial t} - \frac{e}{c}\left[v_y\left\{\frac{\partial A_y}{\partial x} - \frac{\partial A_x}{\partial y}\right\}\right.$$

$$\left. - v_z\left\{\frac{\partial A_x}{\partial z} - \frac{\partial A_z}{\partial x}\right\}\right]. \qquad (A5.1.9)$$

But from the usual relations between fields and potentials in free space we have

$$\left.\begin{aligned}\mathscr{E} &= -\text{grad }\phi - \frac{1}{c}\frac{\partial\mathbf{A}}{\partial t}\\[2mm]\text{and}\qquad\mathscr{H} &= \text{curl }\mathbf{A}\end{aligned}\right\}. \qquad (A5.1.10)$$

Hence from eqn. (A5.1.9) it follows that

$$\frac{d}{dt}\left(p_x + \frac{e}{c}A_x\right) = -e\left(\mathscr{E}_x + \frac{1}{c}\{\mathbf{v}\times\mathscr{H}\}_x\right). \qquad (A5.1.11)$$

On the other hand, from eqn. (A5.1.6) and corresponding equations for v_y and v_z, we obtain, by squaring and adding to find v^2/c^2

$$\left(\mathbf{p} + \frac{e\mathbf{A}}{c}\right) = \frac{m\mathbf{v}}{\left(1 - \frac{v^2}{c^2}\right)^{\frac{1}{2}}}. \qquad (A5.1.12)$$

Substituting this in eqn. (A5.1.11) we have

$$\frac{d}{dt}\left(\frac{mv_x}{\left\{1 - \frac{v^2}{c^2}\right\}^{\frac{1}{2}}}\right) = -e\left(\mathscr{E}_x + \frac{1}{c}\{\mathbf{v}\times\mathscr{H}\}_x\right). \qquad (A5.1.13)$$

But this is simply the x component of the vector equation of motion (A5.1.2) and hence we have justified the Hamiltonian (A5.1.4).

The conclusion is that we introduce the magnetic field into the Hamiltonian by replacing $\mathbf{p}$ by $\mathbf{p} + (e/c)\mathbf{A}$.

APPENDIX 5.2

Momentum wave functions in atoms and wave equation in momentum space

WE have dealt so far with wave functions $\psi(\mathbf{r})$ in coordinate space. However, properties related to electron momenta are observable and we have in mind specifically the shape of the Compton profile in X-ray scattering. Positron annihilation is another experiment which can yield information about electronic moment a (see Jones and March, 1973).

There are two possible approaches to find wave functions in momentum rather than coordinate space. The first is direct solution of the wave equation in momentum space. The second is to use the Dirac transformation theory to pass from the space wave function $\psi(\mathbf{r})$ to the momentum space function $\phi(\mathbf{p})$, via the relations

$$\phi(\mathbf{p}) = \frac{1}{h^{\frac{3}{2}}} \int \exp\left(-i\frac{\mathbf{p} \cdot \mathbf{r}}{\hbar}\right) \psi(\mathbf{r}) \, d\mathbf{r}. \qquad \text{(A5.2.1)}$$

We record also the inverse Fourier transform relation

$$\psi(\mathbf{r}) = \frac{1}{h^{\frac{3}{2}}} \int \exp\left(i\frac{\mathbf{p} \cdot \mathbf{r}}{\hbar}\right) \phi(\mathbf{p}) \, d\mathbf{p}. \qquad \text{(A5.2.2)}$$

Before referring to the use of these relations for direct calculation of momentum wave functions, let us briefly set up the wave equation in momentum space.

(a) *Wave equation in momentum space*

We start from the Schrödinger equation for stationary states

$$H\psi(\mathbf{r}) = E\psi(\mathbf{r}) \tag{A5.2.3}$$

where

$$H = \frac{p^2}{2m} + V(\mathbf{r}). \tag{A5.2.4}$$

We now multiply both sides by $h^{-\frac{3}{2}} \exp\left(-i\dfrac{\mathbf{p}\cdot\mathbf{r}}{\hbar}\right)$ and integrate over all space. We then find

$$\frac{p^2}{2m}\phi(\mathbf{p}) + \frac{1}{h^{\frac{3}{2}}}\int e^{-i\frac{\mathbf{p}\cdot\mathbf{r}}{\hbar}} V(\mathbf{r})\,\psi(\mathbf{r})\,d\mathbf{r} = E\phi(\mathbf{p}). \tag{A5.2.5}$$

This can obviously be rewritten formally as

$$\frac{p^2}{2m}\phi(\mathbf{p}) + \int \mathcal{V}(\mathbf{p}'-\mathbf{p})\,\phi(\mathbf{p}')\,d\mathbf{p}' = E\phi(\mathbf{p}) \tag{A5.2.6}$$

where

$$\mathcal{V}(\mathbf{p}'-\mathbf{p}) = h^{-3}\int V(\mathbf{r}) \exp\left\{\frac{i(\mathbf{p}'-\mathbf{p})\cdot\mathbf{r}}{\hbar}\right\} d\mathbf{r}. \tag{A5.2.7}$$

Thus we have to solve an integral equation in momentum space.

In problems 5.4 and 5.5, you are asked to show that in the case of a harmonic oscillator the integral equation (A5.2.6) reduces to a differential equation, and for the case of a Morse potential it reduces to a difference equation.[†] From the harmonic oscillator example, it is easy to see that the ground-state wave functions in $\mathbf{p}$ and $\mathbf{r}$ space have the same Gaussian form. The Gaussian function is its own Fourier transform and so this is clear from the Dirac transform relation.

[†] For a more detailed discussion of momentum space problems see, for example, the book on Fourier Transforms, by Sneddon (1951a).

Morse, Young and Haurwitz wave functions for heavier atoms

FOR atoms up to Ne, accurate calculations have been made using variational wave functions due to Morse, Young and Haurwitz (1935; see also Duncanson and Coulson, 1944). These are built up from the separate orbitals

$$\psi(1s) = \left(\frac{\mu^3 a^3}{\pi}\right)^{\frac{1}{2}} \exp\left(-\mu a r\right)$$

$$\psi(2s) = (\mu^5/3\pi N)^{\frac{1}{2}} \left[r \exp\left(-\mu r\right) - \left(\frac{3A}{\mu}\right) \exp\left(-b\mu r\right)\right]$$

$$\psi(2p_0) = \left(\frac{\mu^5 c^5}{\pi}\right)^{\frac{1}{2}} r \cos\theta \exp\left(-\mu c r\right)$$

$$\psi(2p_\pm) = \left(\frac{\mu^5 c^5}{2\pi}\right)^{\frac{1}{2}} r \sin\theta \exp\left(-\mu c r \pm i\phi\right) \qquad \text{(A6.1.1)}$$

where r, θ and ϕ are the usual polar angles, N is a normalizing constant, while A is determined by the condition that $\psi(2s)$ and $\psi(1s)$ are orthogonal.

Values of the parameters a, b, c, μ are determined by minimizing the energy for each separate atom and a selection of the results are recorded in Table A6.1.1.

It is then easy to prove that the momentum distribution function $I(p)$ is simply a sum of separate contributions $I_{1s}(p)$, $I_{2s}(p)$, $I_{2p}(p)$, etc.

Self-Consistent Fields in Atoms

TABLE A6.1.1 PARAMETERS IN MORSE–
YOUNG–HAURWITZ ORBITALS
(A6.1.1) FOR LIGHT ATOMS

Atom	a_μ	b_μ	c_μ	μ
Li	2·69	2·26	–	0·658
Be	3·69	3·28	–	0·979
B	4·69	4·20	1·20	1·322
C	5·69	5·13	1·56	1·652
N	6·68	6·08	1·91	1·970
O	7·68	7·06	2·22	2·289
F	8·67	8·02	2·55	2·611
Ne	9·66	8·97	2·88	2·934

Thus the problem is simply that of transforming each separate wave function into momentum space. This is a straightforward generalization of the calculation we have described for the hydrogen $1s$ orbital in section 6.2.

APPENDIX 6.2

X-ray scattering from gas of non-spherical molecules

WE shall briefly deal with the generalization of X-ray scattering from spherical atomic charge clouds to embrace a gas of non-spherical molecules here.

The scattered intensity I_s, in suitable units, is given by

$$I_s = |f|^2 = f^*f$$
$$= \int \varrho(\mathbf{r}_1) \exp\,(ik\mathbf{S}.\mathbf{r}_1)\,d\mathbf{r}_1 \int \varrho(\mathbf{r}_2) \exp\,(-ik\mathbf{S}.\mathbf{r}_2)\,d\mathbf{r}_2$$
$$= \int \int \varrho(\mathbf{r}_1)\,\varrho(\mathbf{r}_2)\,\{\exp\,ik(\mathbf{S}.\mathbf{r}_1 - \mathbf{r}_2)\}\,d\mathbf{r}_1\,d\mathbf{r}_2. \qquad (A6.2.1)$$

This must now be averaged overall orientations which the molecule can take with respect to the incident beam. First of all, as the ϱ's are real, we can write

$$I_s = \int \int \varrho(\mathbf{r}_1)\,\varrho(\mathbf{r}_2)\cos k\mathbf{S}.(\mathbf{r}_1 - \mathbf{r}_2)\,d\mathbf{r}_1\,d\mathbf{r}_2. \qquad (A6.2.2)$$

It is not difficult to show that in order to achieve the desired angular average we can consider the problem as equivalent to allowing $(\mathbf{r}_1 - \mathbf{r}_2)$ to take all possible orientations relative to $\mathbf{S}$. Consider then the factor $\cos k\mathbf{S}.(\mathbf{r}_1 - \mathbf{r}_2)$ and let α be the angle between $\mathbf{S}$ and $\mathbf{r}_1 - \mathbf{r}_2$. The probability that $\mathbf{r}_1 - \mathbf{r}_2$ lies in a direction making an angle between α and $\alpha + d\alpha$ is proportional to the solid angle and is given by

$$\frac{d\omega}{4\pi} = \frac{1}{2}\sin \alpha\,d\alpha. \qquad (A6.2.3)$$

Thus, the average of cos $k\mathbf{S}.(\mathbf{r}_1-\mathbf{r}_2)$ can be written in the form

$$\overline{\cos k\mathbf{S}.\mathbf{r}_1-\mathbf{r}_2} = \int_0^\pi \tfrac{1}{2} \cos\{kSr_{12}\cos\alpha\}\sin\alpha\,d\alpha$$

$$= \sin kSr_{12}/kSr_{12} \qquad (A6.2.4)$$

where $r_{12} \equiv |\mathbf{r}_1-\mathbf{r}_2|$. Thus the scattered intensity takes the form

$$\text{Intensity} = \int\int \varrho(\mathbf{r}_1)\,\varrho(\mathbf{r}_2)\frac{\sin kSr_{12}}{kSr_{12}}\,d\mathbf{r}_1\,d\mathbf{r}_2 \qquad (A6.2.5)$$

where, as can be seen from Fig.6.1.1,

$$S = |\mathbf{s}-\mathbf{s}_0| = 2\sin\frac{\theta}{2}, \qquad (A6.2.6)$$

θ being the scattering angle. It is often convenient to put $kS = K$ and to write

$$f^2(K) = \int\int \varrho(\mathbf{r}_1)\,\varrho(\mathbf{r}_2)\frac{\sin Kr_{12}}{Kr_{12}}\,d\mathbf{r}_1\,d\mathbf{r}_2 \qquad (A6.2.7)$$

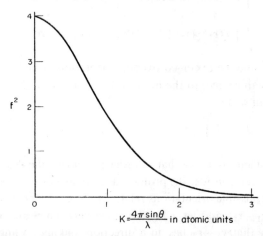

$$K=\frac{4\pi\sin\theta}{\lambda}\ \text{in atomic units}$$

FIG. A6.2.1. Coherent scattering from hydrogen molecule. To graphical accuracy, the scattering from the self-consistent field approximation of Coulson (see Appendix 6.3) the correlated wave function of James and Coolidge (1933) and the Gurnee–Magee wave function (1950) is the same.

Progress can be made by expanding the electron density in the molecule in spherical harmonics (see, Banyard and March, 1957). In some cases, when we insert such an expansion into the above expression for $f^2(K)$, we find rapid convergence, only the first few terms being needed to compute the X-ray scattering.

As a simple example, we show in Fig. A6.2.1 the X-ray scattering from molecular hydrogen, calculated with some simple approximations to the wave function.

APPENDIX 6.3

Self-consistent field for H₂ molecule

COULSON (1938) has calculated an appropriate self-consistent field for molecular hydrogen. He uses coordinates

$$\lambda = \frac{r_a + r_b}{R} \quad \text{and} \quad \mu = \frac{r_a - r_b}{R} \qquad \text{(A6.3.1)}$$

where r_a and r_b are the distances of the electron from the two nuclei A and B, together with the azimuth ϕ around the internuclear axis AB.

He then replaced the Hartree equation by a variation problem, and used trial functions of the form

$$\psi = N \exp\left(-\delta\lambda\right)\left[1 + a\mu^2 + b\lambda + c\lambda^2 + d\lambda\mu^2\right]. \qquad \text{(A6.3.2)}$$

He obtained from his most accurate variational calculations with $R = 1\cdot40$ au, the results

$$N = 0\cdot87758, \quad a = 0\cdot27787, \quad b = -0\cdot12863$$
$$c = 0\cdot012503, \quad d = -0\cdot039589, \qquad \text{(A6.3.3)}$$

δ being chosen as $0\cdot75$ from the outset following the work of James and Coolidge (1933) on H_2^+.

The binding energy thus found was $3\cdot60$ eV, to be compared with the observed value of $4\cdot72$ eV. Coulson's binding energy is certainly fairly near to the best self-consistent-field value for the molecule.

APPENDIX 8.1[†]

Relation between charge density and its gradient at nucleus

THE broad outline of the proof is as follows. An antisymmetrized wave function which has total electron spin quantum numbers S and M can be represented by

$$\Psi_{SM}(r\sigma) = (1/f_{SM})^{\frac{1}{2}} \sum_k \Psi_{Sk}(r)\, \Gamma_{SMk}(\sigma). \qquad \text{(A8.1.1)}$$

Here $\Psi_{Sk}(r)$ denotes the eigenfunctions, which are dependent only on spatial coordinates, common to the $(2S+1)$ functions Ψ_{SM}, while $\Gamma_{SMk}(\sigma)$ denotes the appropriate orthonormalized spin functions. The summation extends over the f_{SM} possible values of k. The first factor on the right-hand side is merely for normalization.

The probability density corresponding to Ψ_{SM} is given by

$$\begin{aligned}\varrho(r) &= \langle \Psi_{SM} | \Psi_{SM} \rangle \\ &= (1/f_{SM}) \sum_k \varrho_k(r) = (1/f_{SM}) \sum_k \sum_n \varrho_k(r_n) \qquad \text{(A8.1.2)}\end{aligned}$$

where

$$\varrho_k(r_n) = \langle \Psi_{Sk} | \Psi_{Sk} \rangle_{r=r_n}, \qquad \text{(A8.1.3)}$$

the integration being over all the space coordinates other than r_n. The summation extends over all n electrons.

---

† The arguments used in this appendix are more advanced than elsewhere in this book.

Self-Consistent Fields in Atoms

Let us denote one of the eigenfunctions Ψ_{Sk} by Ψ. Then it follows from Kato's theorem that

$$\left\langle \Psi \left| \frac{\partial \Psi}{\partial r_n} \right. \right\rangle_{r_n=0} = -Z\varrho_k(0) \qquad (A8.1.4)$$

where $\varrho_k(0)$ is defined from eqn. (A8.1.3) with $r_n = 0$. Using the fact that $\varrho_k(r)$ is real, we have

$$\left(\frac{\partial \varrho_k(r_n)}{\partial r_n} \right)_{r_n=0} = 2\langle \Psi | \partial \overline{\overline{\Psi}} / \partial r_n \rangle_{r_n=0} \qquad (A8.1.5)$$

from which the result

$$\left(\frac{\partial \varrho_k(r_n)}{\partial r_n} \right)_{r_n=0} = -\frac{2Z}{a_0} \varrho_k(0) \qquad (A8.1.6)$$

follows. Therefore we find

$$\left[\frac{\partial \varrho(r)}{\partial r} \right]_{r=0} = -\frac{2Z}{a_0} \varrho(0), \qquad (A8.1.7)$$

this result applying to any eigenstate of an atomic, spin independent, Hamiltonian operator.

APPENDIX 9.1

Solution of Dirac equation for hydrogen atom

WE wish to sketch in this appendix the solution of the radial eqns. (9.21) for the hydrogen atom (see Mott and Sneddon, 1948). Thus we insert the potential energy as (with $Z = 1$ for Hydrogen)

$$V(r) = -Ze^2/r.$$

We shall then consider the solution of the first pair of equations in (9.21), the second pair being solved in essentially the same way.

It is convenient to make the changes of variables given by

$$\zeta = \frac{Ze^2}{\hbar c}, \quad \varepsilon = \frac{E}{mc^2}, \quad \lambda = mc(1-\varepsilon^2)^{\frac{1}{2}}/\hbar, \quad \varrho = 2\lambda r; \quad \text{(A9.1.1)}$$

$$f(r) = i(1-\varepsilon)^{\frac{1}{2}} e^{-\frac{1}{2}\varrho} \varrho^{\gamma-1}\{f_1(\varrho)-f_2(\varrho)\} \quad \text{(A9.1.2)}$$

$$G(r) = (1+\varepsilon)^{\frac{1}{2}} e^{-\frac{1}{2}\varrho} \varrho^{\gamma-1}\{f_1(\varrho)+f_2(\varrho)\}. \quad \text{(A9.1.3)}$$

Then the pair of equations being considered take the form

$$\frac{df_1}{d\varrho} = \left\{1 - \frac{1}{\varrho}\left(\gamma + \frac{\zeta\varepsilon}{(1-\varepsilon^2)^{\frac{1}{2}}}\right)\right\} f_1 + \frac{1}{\varrho}\left\{j + \frac{1}{2} - \frac{\zeta}{(1-\varepsilon^2)^{\frac{1}{2}}}\right\} f_2 \quad \text{(A9.1.4)}$$

and

$$\frac{df_2}{d\varrho} = \frac{1}{\varrho}\left\{j + \frac{1}{2} + \frac{\zeta}{(1-\varepsilon^2)^{\frac{1}{2}}}\right\} f_1 - \frac{1}{\varrho}\left\{\gamma - \frac{\zeta\varepsilon}{(1-\varepsilon^2)^{\frac{1}{2}}}\right\} f_2. \quad \text{(A9.1.5)}$$

Motivated by the non-relativistic case discussed in Appendix 2.1 we assume that the functions $f_1(\varrho)$ and $f_2(\varrho)$ can be represented by the series

$$f_1(\varrho) = \sum_{s=0}^{\infty} c_s \varrho^s ; \qquad f_2(\varrho) = \sum_{s=0}^{\infty} d_s \varrho^s . \qquad (A9.1.6)$$

Substituting in eqns. (A9.1.4) and (A9.1.5) and equating powers of ϱ^{s-1} leads to the recurrence relations

$$\left.\begin{aligned}
c_s \left(\gamma + s + \frac{\zeta \varepsilon}{(1-\varepsilon^2)^{\frac{1}{2}}} \right) - d_s \left(j + \frac{1}{2} - \frac{\zeta}{(1-\varepsilon^2)^{\frac{1}{2}}} \right) &= c_{s-1} \\[2ex]
c_s \left(j + \frac{1}{2} + \frac{\zeta}{(1-\varepsilon^2)^{\frac{1}{2}}} \right) - d_s \left(\gamma + s - \frac{\zeta \varepsilon}{(1-\varepsilon^2)^{\frac{1}{2}}} \right) &= 0
\end{aligned}\right\} \quad (A9.1.7)$$

for the coefficients c_s and d_s. Since we are considering only positive powers of ϱ, we can take $s = 0$, $c_{-1} = 0$, to obtain

$$\left.\begin{aligned}
\left(\gamma + \frac{\zeta \varepsilon}{(1-\varepsilon^2)^{\frac{1}{2}}} \right) c_0 - \left(j + \frac{1}{2} - \frac{\zeta}{(1-\varepsilon^2)^{\frac{1}{2}}} \right) d_0 &= 0 \\[2ex]
\left(j + \frac{1}{2} + \frac{\zeta}{(1-\varepsilon^2)^{\frac{1}{2}}} \right) c_0 - \left(\gamma - \frac{\zeta \varepsilon}{(1-\varepsilon^2)^{\frac{1}{2}}} \right) d_0 &= 0
\end{aligned}\right\} . \quad (A9.1.8)$$

The condition for non-vanishing solutions is then

$$\begin{vmatrix} \gamma + \dfrac{\zeta \varepsilon}{(1-\varepsilon^2)^{\frac{1}{2}}} & -j - \dfrac{1}{2} + \dfrac{\zeta}{(1-\varepsilon^2)^{\frac{1}{2}}} \\[3ex] j + \dfrac{1}{2} + \dfrac{\zeta}{(1-\varepsilon^2)^{\frac{1}{2}}} & -\gamma + \dfrac{\zeta \varepsilon}{(1-\varepsilon^2)^{\frac{1}{2}}} \end{vmatrix} = 0. \quad (A9.1.9)$$

The positive value of γ obtained from this equation is found to be

$$\gamma = \{ (j + \tfrac{1}{2})^2 - \zeta^2 \}^{\frac{1}{2}}, \qquad (A9.1.10)$$

the negative root leading to an unacceptable wave function, since it could not be normalized. Since $j + \frac{1}{2} \geqslant 1$ and $\zeta = Z(e^2/\hbar c) = Z/137 < 1$, it follows that γ is always a real quantity.

If we now eliminate d_s from eqns. (A9.1.7), we get the recurrence formula

$$c_s = \frac{\gamma+s-\zeta\varepsilon/(1-\varepsilon^2)^{\frac{1}{2}}}{(\gamma+s)^2-(j+\frac{1}{2})^2+\zeta^2}\,c_{s-1}. \qquad (A9.1.11)$$

Following again the lines of the non-relativistic treatment, $f_1(\varrho)$ will be a polynomial

$$\sum_{s=0}^{n'-1} c_s\varrho^s$$

in which the last term is non-zero, if

$$\gamma+n'-\frac{\zeta\varepsilon}{(1-\varepsilon^2)^{\frac{1}{2}}} = 0 \qquad (A9.1.12)$$

which can be solved for ε to yield

$$\varepsilon = \pm\frac{\gamma+n'}{\{\zeta^2+(\gamma+n')^2\}^{\frac{1}{2}}}. \qquad (A9.1.13)$$

Adopting the positive root and defining the principal quantum number n by

$$n = n'+j+\tfrac{1}{2} \qquad (A9.1.14)$$

we find the result

$$E = mc^2\left\{1+\frac{\zeta^2}{(n-j-\frac{1}{2}+\gamma)^2}\right\}^{-\frac{1}{2}}. \qquad (A9.1.15)$$

Taking out the rest mass of the electron, and substituting from eqn. (A9.1.10) for γ, the resulting equation is found to be Sommerfeld's fine structure formula

$$E' = mc^2\left[\left\{1+\frac{Z^2e^4/\hbar^2c^2}{\left[n-j-\frac{1}{2}+\{(j+\frac{1}{2})^2-Z^2e^4/\hbar^2c^2\}^{\frac{1}{2}}\right]^2}\right\}^{-\frac{1}{2}}-1\right]. \qquad (A9.1.16)$$

For small values of $Ze^2/\hbar c$, this evidently reduces to the nonrelativistic formula (1.2) for $Z = 1$, or for a general atomic number to

$$E' = -\frac{Z^2}{2n^2}\frac{e^2}{a_0}. \qquad (A9.1.17)$$

References

The numbers in parentheses following each entry refer to the pages in the book where the reference occurs allowing the reference list to be used as an author index.

BAKER, E. B. (1930) *Phys. Rev.* **36**, 630.   (*41*)

BALLINGER, R. A. and MARCH, N. H. (1955) *Phil. Mag.* 46, 246.   (*100*)

BANYARD, K. E. and MARCH, N. H. (1957) *J. Chem. Phys.* **26**, 1416.   (*156*)

BLOCH, F. (1933) *Z. für Physik*, **81**, 363.   (*118, 119*)

BRANDT, W. and LUNDQVIST, S. (1964), *J. Quant. Spectrosc. Radiat. Transfer*, **4**, 679; see also (1963) *Phys. Rev.* **132**, 2135. (*121*)

BRANDT, W., EDER, L. and LUNDQVIST, S. (1967) *J. Quant. Spectrosc. Radiat. Transfer*, **7**, 185, 411.   (*121*)

BROWN, G. E. (1952), *Phil. Mag.* **43**, 467.   (*131*)

BYERS-BROWN, W. and WHITE, R. J. (1970) *J. Chem. Phys.* **53**, 3869.   (*124*)

CONDON, E. U. and SHORTLEY, G. H. (1951) *The Theory of Atomic Spectra* (Cambridge University Press).   (*15*)

COOPER, M. (1971) *Advances in Physics*, **20**, 453.   (*84, 85*)

COULSON, C. A. (1938) *Proc. Camb. Phil. Soc.* **34**, 204.   (*157*)

COULSON, C. A. (1961) *Valence*, 2nd edition (Oxford University Press).   (*15*)

COULSON, C. A. and MARCH, N. H. (1950) *Proc. Phys. Soc.* **A63**, 367.   (*66*)

COULSON, C. A. and NEILSON, A. H. (1961) *Proc. Phys. Soc.* **78**, 831.   (*107*)

DEBYE, P. (1915) *Ann. de Physik*, **46**, 809.   (*77*)

DICKINSON, W. C. (1950) *Phys. Rev.* **80**, 563.   (*50, 57*)

DIRAC, P. A. M. (1928) *Proc. Roy. Soc.* **A117**, 610.   (*126*)

DIRAC, P. A. M. (1930) *Proc. Camb. Phil. Soc.* **26**, 376.   (*94*)

DOYLE, H. T. (1969) *Advances in Atomic and Molecular Physics.* Editors D. R. Bates and I. Eshermann (New York: Academic Press).   (*134*)

DUNCANSON, W. E. and COULSON, C. A. (1944) *Proc. Roy. Soc. Edinburgh*, **62**, 37.   (*152*)

EDLEN, A. (1971) *Topics in Modern Physics.* Editors W. E. Brittin and H. Odabasi (Boulder: Colorado Associated University Press).   (*115*)

EYRING, H., WALTER, J. and KIMBALL, G. E. (1944). *Quantum Chemistry* (New York: Wiley).   (*24*)

FERMI, E. (1928) *Z. für Phys.* **48**, 73.   (*35, 68, 207*)

FEYNMAN, R. P. (1939) *Phys. Rev.* **56**, 340.   (*50, 63*)

FOLDY, L. L. (1951) *Phys. Rev.* **83**, 397.   (*51, 55*)

FREEMAN, A. J. (1959a) *Acta Cryst.* **12**, 261; (1959b) *Acta Cryst.* **12**, 274; (1959c) *Acta Cryst.* **12**, 929. (*83, 87*)

GILVARRY, J. J. (1954) *Phys. Rev.* **95**, 71; **96**, 934. (*132*)

GOMBÁS, P. (1949) *Die Statistiche Theorie des Atoms und Ihre Andwendungen* (Vienna: Springer–Verlag). (*145*)

GOSCINSKI, O. and LINDER, P. (1970) *J. Chem. Phys.* **52**, 2539. (*82*)

GURNEE, E. F. and MAGEE, J. L. (1950) *J. Chem. Phys.* **18**, 142. (*155*)

HARTREE, D. R. (1927) *Proc. Camb. Phil. Soc.* **24**, 89, III. (*31, 169*)

HARTREE, D. R. (1957) *The Calculation of Atomic Structures* (New York: Wiley). (*28, 33, 99*)

HEISENBERG, W. (1931) *Phys. Z.* **32**, 737. (*88*)

HERMAN, F. and SKILLMAN, S. (1963) *Atomic Structure Calculations* (Englewood. Cliffs, N. J.: Prentice–Hall) (*33*)

HOHENBERG, P. and KOHN, W. (1964) *Phys. Rev.* **136**, B864. (*103, 104*)

HULTHÉN, L. (1935) *Z. für Phys.* **95**, 789. (*48*)

HYLLERAAS, E. (1929). *z. für Phys.* **54**, 347; see also ROOTMAN, S.C.J. and WEISS, A. W. (1960) *Rev. Mod. Phys.* **32**, 194. (*106*)

JAMES, H. M. and COOLIDGE A. S. (1933) *J. Chem. Phys.* **1**, 825. (*155, 157*)

JENSEN, H. (1937) *Z. für Phys.* **106**, 620. (*120*)

JENSEN, J. H. D. and LUTTINGER, J. M. (1952) *Phys. Rev.* **86**, 907. (*68, 70*)

JONES, W. and MARCH, N. H. (1973) *Theoretical Solid-State Physics* (London: Wiley). (*111, 149*)

KATO, T. (1957) *Commun. Pure Appl. Math* **10**, 151; (1951) *J. Fac. Sci. Tokyo Univ.* **16**, 145. (*104, 111*)

KELLY, H. P. (1971). *Proc. Menzel Symp. on Solar Physics., Atomic Spectra and Gaseous Nebulae* (Washington DC: Nat. Bur. Stand.) p. 37. (*124*)

KÓNYA, A. (1951) *Acta Physica Hungarica*, **1**, 12. (*66*)

LAMB, W. E. (1941) *Phys. Rev.* **60**, 817. (*57*)

LATTER, R. (1955) *Phys. Rev.* **99**, 150, 1854. (*71, 72, 73, 75*)

LAYZER, D. (1959) *Annals of Physics*, **8**, 271. (*114*)

LAYZER, D. and BAHCALL (1962) *Annals of Physics*, **17**, 177. (*131, 134*)

LESTER, W. A. and KRAUSS, M. (1964) *J. Chem. Phys.* **41**, 1407. (*108, 109*)

LIGHTHILL, M. J. (1958) *Fourier Series and Generalized Functions* (London; Cambridge University Press (*83*)

MARCH, N. H. (1957) *Advances in Physics*, **6**, 1.

MARCH, N. H. and MURRAY, A. M. (1961) *Proc. Roy. Soc.* A**261**, 119. (*104*)

MARCH, N. H. and TOSI, M. P. (1972) *Proc. Roy. Soc.* A**330**, 373. (*118*)

MARCH, N. H. and WHITE, R. T. (1972) *J. Phys. B.* **5**, 466. (*47, 53, 59, 111, 112, 113, 114*)

MARCH, N. H. YOUNG, W. H. and SAMPANTHAR, S. (1967) *The Many–Body Problem in Quantum Mechanics* (Cambridge University Press). (*103*)

MASLEN, V. W. (1956) *Proc. Phys. Soc.* A**69**, 734. (*108, 110*)

McWEENY, R. (1951) *Acta Cryst.* **4**, 513. (*82*)

MILNE, E. A. (1927) *Proc. Camb. Phil. Soc.* **23**, 794. (*46*)

MIRANDA, C. (1934) *Mem. Acc. Italia*, **5**, 283. (*145*)

MORSE, P. M., YOUNG, L. A. and HAURWITZ, E.S. (1935) *Phys. Rev.* **48**, 948. (*67, 152*)

MOTT, N. F. and SNEDDON, I. N. (1948) *Wave Mechanics and its Applications* (Oxford: Clarendon Press). *(134, 160)*

PAULING, L. and WILSON, E. B. (1935) *Introduction to Quantum Mechanics* (New York: McGraw Hill). *(4, 14, 60, 79, 146)*

PIRENNE, M. H. (1946) *The Diffraction of X-rays and Electrons by Free Molecules* (Cambridge University Press). *(88)*

ROOTHAAN, C. C.J., SACHS, L. M. and WEISS, A. W. (1960) *Rev. Mod. Phys.* **32,** 186. *(107)*

RUDKJØBING, M. (1952). *Kgl. Dansbe Videnskab. Selskab. Mat.-Fys. Medd.* **27,** No. 5. *(132)*

RUTHERFORD, D. (1940) *Vector Methods* (Edinburgh: Oliver and Boyd). *(6)*

SCHIFF, L. I. (1955) *Quantum Mechanics*, 2nd edition (New York: McGraw Hill). *(127)*

SCOTT, J. M. C. (1952) *Phil. Mag.* **43,** 859. *(56, 100, 133)*

SEITZ, F. (1940) *Modern Theory of Solids* (New York: McGraw Hill). *(95, 101)*

SHULL, H. and LÖWDIN, P. O. (1956) *J. Chem. Phys.* **25,** 1035. *(64, 124)*

SLATER, J. C. (1933) *J. Chem. Phys.* **1,** 687: (1930) *Phys. Rev.* **36,** 57. *(24, 109)*

SLATER, J. C. (1951) *Phys. Rev.* **81,** 385. *(100, 101, 108, 109, 217)*

SMITH, J. R. (1969) *Phys. Rev.* **181,** 522. *(103)*

SMITH, V. H. (1971) *J. Chem. Phys.* **55,** 482 *(110)*

SNEDDON, I. N. (1951a) *Fourier Transforms* (New York: McGraw Hill). *(150)*

SNEDDON, I. N. (1951b) *Special Functions of Mathematical Physics and Chemistry* (Edinburgh: Oliver and Boyd). *(11)*

STEINER, E. (1963) *J. Chem. Phys.* **39,** 2365. *(105)*

STODDART, J. C. and MARCH, N. H. (1967) *Proc. Roy Soc.* **A299,** 279. *(104)*

THOMAS, L. H. (1926) *Proc. Camb. Phil. Soc.* **23,** 542. *(35, 197)*

VALLARTA, M. S. and ROSEN, N. (1932) *Phys. Rev.* **41,** 708. *(131, 132)*

WALLER, I. and HARTREE, D. R. (1929) *Proc. Roy. Soc.* **A124,** 119. *(88)*

WEISS, A. W. (1961) *Phys. Rev.* **122,** 1326. *(110)*

WENDIN, G. (1972) *J. Phys. B.* **5,** 110, and other references given there. *(123)*

WIGNER, E. P. (1934) *Phys. Rev.* **46,** 1002. *(103, 124)*

WIGNER, E. P. (1938) *Trans. Faraday Soc.* **34,** 678. *(103, 124)*

WIGNER, E. P. and SEITZ, F. (1934) *Phys. Rev.* **46,** 509. *(100, 102)*

WILLIAMS, A. O. (1940). *Phys. Rev.* **58,** 723. *(130)*

WILSON, W. S. and LINDSAY, R. B. (1935) *Phys. Rev.* **47,** 681. *(30)*

PART II

Notes to papers reprinted below

1. Though Hartree's paper reprinted here is Part II of a series, it should be readily understood with the background gained from the main text. This appears to be the origin of the 'self-consistent' field terminology.

2. Thomas's paper is reprinted in full, but the table on the last page is obsolete, and use must be made of the table in Appendix 3.1 instead.

3. The paper by Fermi is translated from the German by D. H. Webb

4. Slater's paper is reprinted, rather than the original papers on the Hartree–Fock equations, because it gives:

   (a) A clear physical interpretation of these equations.
   (b) An important simplification following the lines of Dirac (1930).

1

The Wave Mechanics of an Atom with a Non-Coulomb Central Field. Part II. Some Results and Discussion

D. R. HATREE (*Proc. Camb. Phil. Soc.* **24**, p. 111–132)

St John's College

[*Received* 19 November, *read* 21 November 1927.]

§ 1. Introduction

In the previous paper an account has been given of the theory and methods used for determining the characteristic values and functions of Schrödinger's wave equations for a non-Coulomb central field of force whose potential v is given. In this paper some results of the application of these methods will be given and discussed; references to the previous paper will be prefixed by a I. Atomic units (see I, § 1) will be used throughout this paper*.

There are two different objects which one may have in view in doing numerical work of this kind.

First, the object may be to find an empirical field of force for a given atom, for which the characteristic values of the wave equation give as closely as possible the terms of the optical and X-ray spectra of that atom.

Secondly, as suggested in I, § 1, it may be to find a field of force such

* *Editor's note:* i.e. $e = m = h = 1$.

M-sfa 12

that the distribution of charge given by the wave functions for the core electrons shall reproduce the field*.

Certainly these are not altogether independent; in the first case, one may hope that the empirical field may be expressible as the sum of contributions from the different core electrons[†], and in the second case, one may hope that the characteristic values in the field so found will give approximately the term values of the spectra of that atom. Nevertheless they are distinct in that in one case the atomic field may be chosen without reference to the characteristic functions themselves, and in the other case depends essentially on them, so that they involve rather different procedure in the numerical work. The first would be the more appropriate if the results were to be used to find the relative intensities of lines in the spectrum, and some work from this point of view has already been done by Sugiura*, the second, if the distribution of charge were required[†] and further results were to be based on it (for example X-ray scattering factors F).

The work here described was done with the second as its main object, and this demands consideration of the appropriate field of force to use for the core electrons.

§ 2. The Field of Force for a Core Electron

Consider an atom such as the neutral atom of an alkali metal, consisting of closed n_k groups and a series electron. The potential v for the

---

* Corresponding work for the orbital atomic model has been done from the first point of view by E. Fues, *Zeit. f. Phys.*, Vol. XI, p. 364 (1922); Vol. XII, p. 1.; Vol. XIII, p. 211 (1923); Vol. XXI, p. 265 (1924); D. R. Hartree, *Proc. Camb. Phil. Soc.*, Vol. XXI, p. 265 (1923) and Y. Sugiura and H. C. Urey, *Det Kongel. Danske Videnskab. Selskab., Math.-Phys. Medd.*, Vol. VII, No. 13 (1926), and from the second by R. B. Lindsay, *Publ. Mass. Inst. Technology*, Series II, No. 20 (1924).

† For the orbital atomic model the writer has tried to obtain such results, but without success.

* Y. Sugiura, *Phil. Mag.*, Ser 7, Vol. IV, p. 495 (1927).

† The determination of the distribution of charge directly from a potential or field given as a function of the radius involves numerical differentiation, which is an unsatisfactory process, especially in this case, when it is possible to make a small increase of the field at one radius and a small decrease at another without appreciably affecting the fit between calculated and observed term values.

series electron is that of the field of the centrally symmetrical distribution of charge of the closed groups; but just as in the theory of the hydrogen atom the field acting on the electron is that of the nucleus only, not that of the nucleus and its own distributed charge, so here the field for a core electron is the total field of the nucleus and all the closed groups, less its own contribution to that field. Now except for an electron with $l = 0$, its own contribution to the field is not centrally symmetrical, so that it would seem that the assumption of a central field is not applicable to it; on the other hand, the structure of the X-ray terms is that of terms due to a spinning electron in a central field.

It is just here that we meet the most serious doubts concerning the replacement of the actual many-body problem by a one-body problem with a central field for each electron, even as a first approximation.

A doublet term of an X-ray spectrum is due not to the presence of a single definite electron, but to the absence of one from an otherwise complete n_k group; the presence of this last electron gives the single ^1S term characteristic of a complete group. Now it is only through operation of Pauli's exclusion principle[*] that an n_k group lacking one electron to make it complete gives only a spectral term of the same type (multiplicity and *l*-value) as would be given by that electron alone in a central field of other complete groups, and that the complete group gives only a ^1S term; this principle is essentially an expression of the behaviour of a number of electrons with the same n_k, whose mutual interactions may be considerable, and this involves developments of quantum mechanics beyond those applied in this paper.

In order to make headway at all without departing from the simple idea of a central field for each electron, a simplifying assumption is necessary here in dealing with core electrons, and in this paper it will be assumed that the appropriate potential to take in working out the characteristic value and function for a core electron is the total potential of the nucleus and of the whole electronic distribution of charge, less the potential of the centrally symmetrical field calculated from the distribution of charge of that electron, *averaged over the sphere for*

[*] See, for example, F. Hund, *Linienspektren*, p. 114 *et seq.*

12*

each radius; the radial density of this averaged distribution of charge is just the normalised value of P^2 $\left(\text{i.e. the charge between radii } r \text{ and } r+dr \text{ is } P^2 \, dr \middle/ \int_0^\infty P^2 \, dr\right)$ or, for a complete n_k group consisting of s electrons, it is $1/s$ of the contribution to the radial density from this whole group.

It would of course be possible in most cases to carry out the work for the core electrons without taking into account at all the fact that the distributed charge of a core electron does not contribute to the field acting on it, but it seems probable that a better approximation to the actual distribution of charge will be obtained by making some correction, even if it is only a crude one. Besides, in treating the outer electrons of negative ions, some such correction is necessary in order to make the work possible at all.

For numerical work we have to start from a field which will be called the 'initial field'; for each n_k corresponding to a group of core electrons the field is corrected, as explained above, for the fact that the distributed charge of an electron must be omitted in finding the field acting on it, and for the field so corrected the part of the solution of the wave equation depending on r is found by the methods given in I; then from the solutions for all groups of core electrons a distribution of charge can be calculated (if the n_k groups are all complete, this distribution of charge will be centrally symmetrical), and then the field of the nucleus and this distribution of charge can be found; this may be called the 'final field.' The process may be expressed briefly in diagrammatic form:

Initial Field
↓
Initial Field corrected for each core electron
↓
Solutions of Wave Equation for core electrons
↓
Distribution of Charge
↓
Final Field.

If the final field is the same as the initial field, the field will be called 'self-consistent,' and the determination of self-consistent fields for various atoms is the main object of this paper. The self-consistent field so found is a characteristic of the particular atom in the particular state of ionisation considered; it involves no arbitrary functions or constants whatever.

It is thought that the distribution of charge in the self-consistent field is probably the best approximation to the actual distribution of charge in the atom which can be obtained without very much more elaborate theoretical and numerical work, and so is the most suitable to use in any problems involving this distribution of charge; also it is hoped that when the time is ripe for the practical evaluation of the exact solution of the many-electron problem, the self-consistent fields calculated by the methods given here may be helpful as providing first approximations.

For each solution of the wave equation in the self-consistent field (corrected as already explained) in the case of a core electron there will be a characteristic value of the energy parameter ε; it is a further question whether these values of ε are directly related to the optical and X-ray term values by the relation

$$\varepsilon = \nu/R.$$

For the core electrons, this again depends on the details of the mutual interaction of the electrons in a closed group, and on the effect of the removal of one on the remainder; the simplest case, that of the normal state of neutral helium, has been worked out, and suggests that this relation may be expected to hold closely for the X-ray terms (see § 4).

§ 3. Some Practical Details

The determination of the self-consistent field for any atom is a matter of successive approximation. Fortunately this process is 'stable' in that change of the initial field, of the same sign throughout, gives a change of the final field of the opposite sign. Consider, for example, an increase in the initial (positive) field; this pulls the whole distribution of charge further in, so that the negative charge inside a given radius is greater and the final field at each radius less than it was without the increase in the initial field.

For the first atom tried the initial field for the first approximation was obtained by Thomas' method*; in this case preliminary solutions of the wave equations have to be found with the same initial field for all core electrons, without the correction referred to above, in order to estimate the correction. It is better to build up an initial field as the sum of estimated contributions from the different groups of core electrons, so that the corrected field for each can be used even in the first approximation; when the calculations for two or three atoms have been completed this building up of the initial field can be done quite satisfactorily by interpolation or extrapolation.

Since a given change in Z means a smaller *proportional* change in the field at small radii than at large radii, it might be expected that the final field would be more sensitive to changes in Z of the initial field at large radii than at small. This proves to be the case, and, with the stability of the process of successive approximation to a self-consistent field, suggests that the initial field at any stage should be taken between the initial and final fields of the previous approximation, nearer the final field at small r and nearer the initial field at large r. This general rule is justified by experience[†].

For a central field of potential v, which is a function of r only,

$$Z = -r^2 \, dv/dr = dv/d(1/r)$$

is the point charge which, placed at the nucleus, would give the same *field* at radius r as the actual field; it is often called the 'effective nuclear charge' and this name will be used here. The quantity

$$Z_p = rv$$

---

* L. H. Thomas, *Proc. Camb. Phil. Soc.*, Vol. XXIII, p. 542 (1927). It is necessary to extrapolate the field empirically beyond the range to which Thomas' results apply.

† The process of successive approximation by taking the initial field always equal to the final field of the previous approximation is not always convergent, though perhaps it usually is. Even when it is, a more rapid convergence to the self-consistent field is obtained by the rule given here.

is the point charge, which, placed at the nucleus, would give the same *potential* at radius r as that of the actual field; it may be called the 'effective nuclear charge for potential*.' It is usually convenient to work with Z and Z_p rather than with the field and the potential themselves, since Z and Z_p vary over a much smaller range. Differentiating $Z_p/r = v$ and substituting for dv/dr we have

$$-\frac{dZ_p}{dr} = \frac{Z-Z_p}{r},$$

and when Z is given as a function of r it has proved more satisfactory to find Z_p by numerical integration of this equation (which is very easy) than by direct integration of $v = -\int Zr^{-2}\, dr^\dagger$.

In the numerical work, two places of decimals have usually been kept in Z in the initial and final fields, and the successive approximations towards a self-consistent field have been continued until the maximum difference in Z between the initial and final fields became less than 0·1.

§ 4. The Normal State of Neutral Helium

The simplest case of an atom with more than one electron to which the idea of the self-consistent field is applicable is that of the normal state of neutral helium. In this case we consider the motion of one electron in the field of the nucleus and the second electron, and try to determine this field so that the 1_1 wave function for the first electron gives a charge distribution for which the field is the same as that assumed for the second electron.

---

* The large difference between Z and Z_p is not always realised; for Rb the maximum value of Z/Z_p is over 2 and that of $Z-Z_p$ is over 13, so that Z and Z_p cannot be considered as even approximately equal for a non-Coulomb field.

† The reason is that Z is known at equal intervals of r, and for the direct integration unduly small intervals would have to be used in some regions, in order to keep down the higher orders of difference due to the r^{-2} factor; the formula for the mean value $\bar{Z}$ for $\delta v = \bar{Z}\delta(1/r)$, when Z is known at equal intervals of r not of $1/r$, is more trouble to use than the differential equation.

The numerical work went so easily that an extra decimal place was kept in Z, and the successive approximation was carried to a point where the maximum difference between the values of Z for the initial and final fields was 0·002.

The value of Z and the radial charge density $-dZ/dr$ for this field are given in Table I.

TABLE I NEUTRAL HELIUM, NORMAL STATE. SELF-CONSISTENT FIELD
AND DISTRIBUTION OF CHARGE

r atomic units	Z	$-dZ/dr$ electrons per atomic unit	r atomic units	Z	$-dZ/dr$ electrons per atomic unit
0	2·000	0·00	1·6	0·239	0·48
0·1	1·988	0·30	1·8	0·159	0·33
0·2	1·932	0·83	2·0	0·105	0·22
0·3	1·826	1·28	2·2	0·068	0·15
0·4	1·682	1·57	2·4	0·044	0·10
			2·6	0·028	0·06
0·6	1·344	1·73	2·8	0·018	0·04
0·8	1·013	1·55	3·0	0·011	0·026
1·0	0·733	1·25			
1·2	0·515	0·94	3·5	0·003	0·009
1·4	0·354	0·68	4·0	0·001	0·003

The most interesting point about the calculation is the value of the energy parameter ε, which was found to be $\varepsilon = 1·835$ (24·85 volts), while the observed term value for the normal state of neutral helium gives $v/R = 1·81$ (24·6 volts). Now the calculated value of ε is found by considering the wave equation for one electron in the field of the nucleus and the distributed charge of the other, without taking into account the effect of the removal of one on the energy or distribution of charge of the other, while in the actual removal of one the change of charge distribution of the other is quite appreciable. Until it can be shown that the calculated energy parameter of the wave equation, for one electron in the self-consistent field of the nucleus and the other, would be expected to agree well with the energy required to remove one electron, the other adjusting its charge distribution to the change of field caused by that removal, the very close agreement actually found must be taken as an empirical justification of the simple approximations used to represent the interactions of electrons with the same n_k, and of the general idea underlying the work, rather than as a notable success of the method. But it does suggest that in general for core electrons the energy parameter of the wave equation for the self-consistent field, corrected as already explained, may be expected to be a good approximation to the value of ν/R for the corresponding X-ray term*.

§ 5. Rubidium

These methods were first applied to the Rb atom, and more extensive work has been done for it than for any other atom. This choice was made because it seemed interesting to examine what happened on the wave mechanics when on the orbital model an internal and an external orbit of the same n_k were possible, and this does not occur for elements of too low atomic number [Cu(N = 29) is the first such atom which also gives an optical spectrum which can be dealt with by a central field]; for Rb it does occur, and at the same time the atomic number is small enough for the effects of the relativity variation of mass and of the spinning electron to be dealt with adequately as independent first order perturbations (except perhaps for $l = 0$); also for Rb, the writer had fairly extensive results worked out for the orbital model, and it seemed likely that these would be interesting for comparison.

The work for the core electrons was carried out for the ion Rb+. For the first approximation the initial field was calculated by Thomas' method already mentioned, without the correction for the fact that a core electron is not acted on by its own distribution of charge; the

---

* Actually the work was done in the inverse order; the application of the method to helium as an experiment was suggested by the good agreement between values of ε so calculated and the observed X-ray term values for more complicated atoms.

final field for the first approximation was taken as the initial field for the second, and the correction was applied. Some trial changes in the field were then made to give an idea of the sensitivity of the final field to a change in the initial field, and a third initial field was built up from estimates of the contributions to Z from the different groups of core electrons. The maximum difference between this initial field and the final field calculated from it was $0 \cdot 08$ in Z; this was within the limit of disagreement ($0 \cdot 1$ in Z) which had been previously laid down, so that the process of approximation was not carried further. Fig. 1*

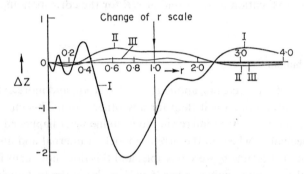

FIG. 1. Showing successive approximations to self-consistent field. Difference ΔZ between effective nuclear charge of initial and final field plotted against r for the three approximations (Curves I, II, III) made in the calculations for Rb.

shows the process of successive approximation to a self-consistent field; for the three approximations the difference between the values of Z for the initial and the final field is plotted as a function of r.

The effective nuclear charge Z and the radial charge density $-dZ/dr$ of the final field[†] of the third approximation is given in Table II: the

---

* In Figs. 1, 2, 3 a different scale of r is used for $r < 1$ and $r > 1$. A scale open enough to show the detail of the curves for large r is unnecessarily open for larger r; the use of two uniform scales has seemed preferable to the continuous distortion introduced, for example, by a logarithmic scale of r.

† The charge density is calculated from the wave functions from which the final field is built up, not by numerical differentiation of Z.

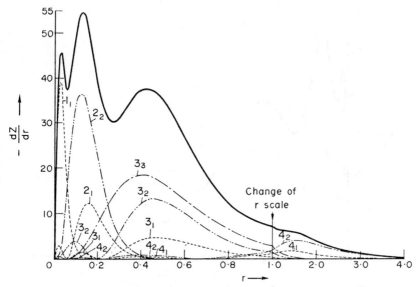

FIG. 2. Radial distribution of charge for Rb^+ and contributions from different groups of core electrons. Radial density $-(dZ/dr)$ in electrons per atomic unit plotted against r in atomic units. ——— Total for all core electrons ········· Contributions from groups with $l = 0$ $(k = 1)$ — ·· — ·· — Contributions from groups with $l = 1$ $(k = 2)$ ——–——– Contributions from groups with $l = 2$ $(k = 3)$

values of Z probably differ from those of the self-consistent field by less than 0·05 throughout.

§ 6. Rubidium. Comparison of Charge Distribution Calculated by Different Methods

In Fig. 2* the separate contributions to the radial charge density from the different groups of core electrons are shown, also the total radial density; the regions of maximum charge density corresponding to the K, L, M 'shells' will be noticed, but there is no peak, only a flattening out of the curve, for the N 'shell', as the contribution from the M electrons is still appreciable and decreases more quickly than

* See footnote on p. 178.

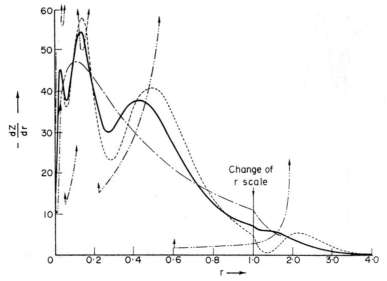

FIG. 3. Radial distribution of charge for Rb⁺ calculated by various
methods.

Radial density $-(dZ/dr)$ in electrons per atomic unit plotted as a
function of radius r in atomic units.

Curve I ————— calculated by method of self-consistent field as
described in the present paper

Curve II — — — — by Thomas' general method

Curve III — ·· — ·· from orbital atomic model (half integer values
of k; approximate only)

Curve IV ············ by Pauling's method

the contribution from the N electrons increases. In Fig. 3 the charge
distribution curve (Curve I) thus calculated is compared with those
calculated by other methods, viz.:

(1) By the general method of Thomas* (Curve II). In effect, Thomas'
work deals with the solution in classical mechanics of the problem

* *loc. cit.*

investigated here on the basis of the wave mechanics. In the notation of this paper, Thomas' equation (1.2) gives

$$-dZ/dr = (8\sqrt{2}/3\pi)r^2v^{\frac{3}{2}} = (8\sqrt{2}/3\pi)\sqrt{(rZ_p^3)}.$$

The potential v (or preferably $Z_p = rv$, which varies more slowly and thus is easier to interpolate) can be calculated from the table at the end of Thomas' paper and the radial charge density follows directly from this equation.

TABLE II RUBIDIUM⁺. APPROXIMATE SELF-CONSISTENT FIELD AND DISTRIBUTION OF CHARGE.

r atomic units	Z	$-dZ/dr$ electrons per atomic unit	r atomic units	Z	$-dZ/dr$ electrons per atomic unit
0	37·00	0	0·40	21·56	37·3
0·005	36·99	7·6	0·45	19·68	37·5
0·01	36·92	21·0	0·50	17·85	35·2
0·02	36·60	40·8	0·6	14·71	26·9
0·03	36·16	45·6	0·7	12·45	18·5
0·04	35·71	42·6	0·8	10·93	12·5
0·05	35·30	39·0	0·9	9·89	8·9
0·06	34·92	37·5	1·0	9·09	7·05
0·07	34·53	38·9			
0·08	34·14	41·8	1·2	7·81	6·08
			1·4	6·60	5·90
0·10	33·23	49·3	1·6	5·44	5·47
0·12	32·20	54·0	1·8	4·43	4·69
0·14	31·12	54·1	2·0	3·60	3·82
0·16	30·06	50·2			
0·18	29·11	44·6	2·5	2·21	2·08
0·20	28·29	39·2	3·0	1·52	0·90
			3·5	1·21	0·38
0·25	26·52	31·0	4·0	1·09	0·16
0·30	25·02	31·2	4·5	1·04	0·06
0·35	23·36	34·8	5·0	1·01	0·02

(2) From the orbital atomic model, with half integer values of k (Curve III). This curve is approximate only*.

(3) From the approximate application of the wave mechanics suggested by Pauling[†] (Curve IV), in which the radial distribution of charge for an electron is taken to be that for an electron with the same n_k in a certain Coulomb field, the nuclear charge for this Coulomb field, different for each group of core electrons, being given by an empirical table.

It will be seen that the distribution given by Thomas' method is a good smoothed-out approximation to the distribution giving the self-consistent field; the curves intersect several times, as do also the curves for Z, the integrated charge density, as can be seen roughly from the curve in Fig. 1 for the first approximation, for which the initial field was calculated by Thomas' method. The maximum difference in Z between the self-consistent field and Thomas' field is just less than 2.

If the details of the field or charge distribution are unimportant for any application, it seems very possible that Thomas' field or distribution might very well be accurate enough. If it is used, however, one point should be kept in mind, namely that it gives a radial charge mechanics the for small r is proportional to $r^{\frac{1}{2}}$ while on the wave density which radial density for small r is proportional to r^2; this may cause apparently large divergences in results involving, for example, $r^{-1} dZ/dr$, such as arise in calculating the charge distribution in plane sheets.

A glance at Fig. 3 shows that the charge distribution calculated from the wave mechanics and from the orbit model of the atom are entirely different in character. For the orbit model the contribution from a group of core orbits to the radial density near the outer apse, of radius r_0, becomes infinite like $(r_0-r)^{\frac{1}{2}}$ as $r \to r_0$ from below, and

---

* For simplicity it has been assumed that all orbits of the same n have the same inner apsidal distance and that all of the same k have the same outer apsidal distance, bothof which assumptions are approximately but not accurately true. The orbital model has not been worked out completely for half integer values of k, but the curve is drawn from data estimated from the work with integer values.

† L. Pauling, *Proc. Roy. Soc.*, Vol. CXIV, p. 181 (1927).

falls discontinuously to zero as r passes through r_0, and there is similar distribution outside the inner apse, so that the radial density curve for the whole atom has infinite discontinuities at the apsidal distances of all the orbits*. The distribution according to the wave mechanics remains always finite and continuous; a particular point to notice about it is that it tends asymptotically to zero (roughly exponentially) as r increases, so that it is not possible to assign a definite size to the atom. Another point is that the successive maxima in the radial density curve (for $n > l+1$) calculated by the wave mechanics have no analogy whatever on the orbital model.

For small radii the distribution calculated by Pauling's method agrees well with that calculated from the self-consistent field, as would be expected, since Pauling's method is based on the wave functions in a Coulomb field, and for small radii the main contribution to the radial density is from electrons for which the proportional deviation of the field from a Coulomb field is small over the range where the wave function is appreciable. For large radii, however, the difference between the two distributions increases, till for the outermost electrons there is little agreement between them.

There are two reasons for this. First, the scale of the distribution of charge as given by Pauling is too large (i.e. the 'size screening constant' given by Pauling's Table VIII is too large†), the error in scale increasing with n (the principal quantum number) and being quite large for the outermost groups of core electrons ($n = 4$ in this case). Secondly, Pauling takes the radial distribution for any one group of electrons to be proportional to that of an electron in a certain Coulomb field; the actual deviation from Coulomb field is always such that the effective nuclear charge decreases with increasing r, and the effect of this is to make the peaks of the radial density curve lower and less sharp than they are for a Coulomb field, and particularly to decrease the rate at which the radial density beyond the outermost

---

* With the half integer values of k there are no circular orbits. The curve for the orbital model is really much more broken than as shown in Fig. 3.

† The effect of Pauling's correcting factor ΔS_s is small in this case, so that the errors lie in the values of S_{s_0}.

maximum falls off with increasing r. These two effects combine to make the maxima and minima of the radial distribution curve calculated by Pauling's method too pronounced; in particular with Pauling's distribution of charge the contribution to the radial density from the M electrons has become small before that due to the N electrons has begun to increase up to its outermost maximum, so that the radial density drops to a very small value at about $r = 1.3$; with the self-consistent distribution of charge the contribution from the M electrons falls off more slowly, and the increase of that from the M electrons begins at a smaller radius, so that there is actually no minimum between the M and N 'shells'.

The comparison between the distribution of charge calculated in different ways has been based on the results for Rb, but, with appropriate alterations in details, most of it would apply to any atom.

§ 7. Rubidium. Comparison of Calculated and Observed Term Values

The characteristic values of the energy parameter ε for the different solutions of the wave equation in the self-consistent field may be compared with the terms of the X-ray and optical spectra.

We consider first the X-ray terms; for these, the removal of one electron is presumably accompanied by some readjustment of the remainder of the atom, but the results for helium suggest that without taking this into account, the values of ε so calculated from the self-consistent field (corrected for the fact that the distributed charge of a core electron does not contribute to the field on itself) may be in good agreement with the X-ray term values. A comparison is given in Table III; the effects of the relativity variations of mass and of the spinning electron were calculated as first order perturbations by the formulae given in I, § 8. The agreement is strikingly good, especially when it is remembered that there is no arbitrary function or constant available to be adjusted to bring calculated results into agreement with the observations. The calculated term values refer to the Rb⁺ ion; those for the neutral atom would probably be smaller by about 0·4 on

TABLE III RUBIDIUM. COMPARISON OF OBSERVED X-RAY TERM VALUES AND ENERGY PARAMETERS FOR CORE ELECTRONS.

n_k		ν/R obs.*	ε calc. for Rb$^+$ ion				ν/R obs. $-\varepsilon$ calc.	ν/R obs. $-\varepsilon$ calc. for orbit model
			Final approximation to self-consistent field	Spin Correction	Relativity Correction	Total		
1_1	K	1119·1	1103	−100	+113	1116	−3	0
2_1	L_{I}	152·3	144·5	−9·8	+15·1	149·8	−2·4	−0·6
2_2 $\begin{cases} L_{\mathrm{II}} \\ L_{\mathrm{III}} \end{cases}$		137·6 \ 133·2	132·3	+2·8 \ −1·4	+2·4	137·5 \ 133·3	−0·1 \ +0·1	0
3_1	M_{I}	24·1	21·24	−1·56	+2·51	22·19	−1·9	−0·0
3_2 $\begin{cases} M_{\mathrm{II}} \\ M_{\mathrm{III}} \end{cases}$		18·2 \ 17·6	16·55	+0·44 \ −0·22	+0·44	17·43 \ 16·77	−0·8 \ −0·8	−0·1
3_3 $\begin{cases} M_{\mathrm{IV}} \\ M_{\mathrm{V}} \end{cases}$		8·4 unresolved	8·28	+0·12 \ −0·08	+0·08	8·48 \ 8·28	(0)	0
4_1	N_{I}	(2·4)	2·964	−0·198	+0·322	3·088	(+0·7)	
4_2 $\begin{cases} N_{\mathrm{II}} \\ N_{\mathrm{III}} \end{cases}$		(1·3)	1·557	+0·040 \ −0·020	+0·041	1·638 \ 1·578	(+0·3)	

* The observed values for all terms but the K term have been taken from a paper by D. Coster and F. P. Mulder, *Zeit. f. Phys.*, Vol. XXXVIII, p. 264 (1926). The value for the K term is from M. Siegbahn, *Spectroscopy of X-rays*.

account of the effect of the distributed charge of the series electron in decreasing the potential inside the core; the field and so the distribution of charge of the core electrons would not be altered appreciably. The observed 4_1 and 4_2 terms are probably somewhat uncertain apart from this effect, which is relatively largest for them.

For comparison, the differences between the observed and calculated term values on the orbital model are also given, the field used in these calculations being chosen to give as good a fit as possible. Compared with these differences, those between term values observed and calculated by the methods of this paper look rather large, but it must be remembered the latter calculated values depend on the self-consistent field and do not involve even one adjustable constant, so that they could be calculated without knowledge of the observed term values, while for the orbit model the calculated values depend on a whole adjustable function, the field, which is chosen empirically to give the best fit to the observed terms. When we come to consider the optical term, it will be seen that even so the advantage lies with the wave mechanics.

The agreement between the calculated and observed spin doublets is very satisfactory; for the L spin doublet the separation deduced from direct observations is 4·4, the value calculated from the general empirical (first order) formula $(\alpha^2/16)\,(N-3\cdot5)^4$ is 4·19, and the value calculated by the perturbation formula (also first order) is 4·20. The calculated M doublet separation agrees with that observed within the limits of observational error. It seems possible that first order relativity and spin corrections may not be adequate for the terms with $l = 0\,(k = 1)$, for which the first order corrections separately are about 10 per cent. of the total term value; for these terms also the difference between observed and calculated term values is largest.

As an example of the magnitude of the correction for the fact that the distributed charge of a core electron does not contribute to the field acting on itself, it may be mentioned that for the 2_2 term this correction is about 12, so that if it were not taken into account the agreement between observed and calculated term values would be entirely lost.

For the optical terms, the quantum defect $q = n - n^*$, i.e. the difference between the principal and effective quantum numbers, is the
most suitable quantity to use in making a comparison of the results of
calculation with observation. Such a comparison is given in Table IV,

TABLE IV RUBIDIUM. COMPARISON OF OBSERVED AND CALCULATED
VALUES OF QUANTUM DEFECT FOR OPTICAL TERMS.

n_k	$q = n^* - n$ obs.	q calculated				q obs. $-q$ calc.	q obs. $-q$ calc. orbit model
		Calc. final approx.	Spin Correction	Relativity Correction	Total		
5_1	3·195	2·986	−·034	·056	3·008	+·187	−·224
6_1	3·153	2·964	−·034	·057	2·987	·166	−·160
7_1	3·146	2·960	−·034	·057	2·983	·163	−·130
5_2	{ 2·707 / 2·720	2·519	−·007 / +·014	·015	2·527 / 2·548	·180 / ·172	−·010
6_2	{ 2·670 / 2·683	2·494	−·007 / +·014	·015	2·502 / 2·523	·168 / ·160	+·058
4_3	{ 0·233 / $\Delta v = (7)$	0·028	Doublet $\Delta v = 3·3$	·000	0·028	·205	

in which the calculated values are for the final approximation to a self-
consistent field, the observed values are calculated from the terms given
in Fowler's table[*], and for comparison the differences between observed and calculated values of the quantum defect are given for the orbital atomic model, with the field found to give about the best general
agreement between the calculated and observed optical and X-ray
terms[†]. The differences of quantum defect between observation and

[*] A. Fowler, *Report on the Series in Line Spectra*, p. 104. The first d term has
been given the principal quantum number $n = 4$ in accordance with § 8 of the present
paper. The doublet separation for this term has not been observed and the value
given is estimated from the value observed for the second d term.

[†] Integer values of k were used in these calculations. The writer has tried some
work with half integral values of k, but without any very marked improvement in
the general agreement between observed and calculated values.

calculation on the wave mechanics appear at first sight rather large, but they are all positive and of about the same magnitude, which is very satisfactory as will be seen shortly. The corresponding differences on the orbit model show a very much wider range of variation and the largest is greater than the largest with the wave mechanics; moreover, this is the best agreement attained with a whole arbitrary function available for adjustment.

It was one of the unsatisfactory points about the results for the orbit model that the calculated variation of the quantum defect within a sequence of terms of the same k (i.e. the deviation from a simple Rydberg formula) was very much larger than the observed variation (larger by a factor from 2 to 3 for the s terms and from $1\frac{1}{2}$ to $2\frac{1}{2}$ for p terms, the error for the first term being relatively the largest). This was found to be the case for all atoms for which calculation was done and could not be avoided by using half-integer values of k. For the results according to the wave mechanics considered here the calculated variation is actually less than the observed, and this is satisfactory as will appear shortly.

The reasons why the differences between the observed quantum defect and that calculated on the wave mechanics are more satisfactory than they appear at first sight is this. The series electron when outside the core presumably polarises it, and the resultant polarisation gives rise to an attractive force on the series electron, always central, in addition to the field due to the unperturbed distribution of charge*. The additional field would act on the series electron as a central perturbing field having an inverse fourth power potential at large distances. It would certainly increase all calculated values of the quantum defect, and from the effect of such a perturbing field on the orbit model it may be expected that for the s and p terms it would increase the calculated quantum defect somewhat more for the first term or two of each

* In the case of He (terms other than s terms) the core consists of a hydrogenlike system, for which the second order Stark effect, on which the polarisability depends, can be worked out exactly for a uniform perturbing field. This has already been treated on the wave mechanics by I. Waller, *Zeit. f. Phys.*, Vol. XXXVIII, p. 635 (1926); see particularly § 3.

sequence than for the rest, and would affect the two sequences about equally; this is just what is required to improve the agreement between calculated and observed values, both for the individual terms and for the variation of quantum defect within each sequence. How the d terms would be affected is not so easy to say.

This effect has not been calculated as it is not yet clear how the perturbing field may be expected to vary when the distance of the series electron from the core is not large compared with the dimensions of the latter, and this is the important part of the range for application to these terms. It would seem necessary at present to use an empirical perturbing field with a potential proportional to r^{-4} for large r, the behaviour for small r being chosen to give—in combination with the self-consistent field due to the unperturbed core—the best fit for the observed terms.

One unsatisfactory point about the results for the optical terms is the magnitude of the p spin doublet, the calculated separation being about $1\frac{1}{2}$ times that observed. A numerical mistake is a possible explanation, but the work has been thoroughly checked and it seems very unlikely that a mistake anything like big enough to explain the discrepancy could have escaped detection. Another possible explanation is this. The magnitude of the spin doublet on the new quantum mechanics was worked out by Heisenberg and Jordan for a Coulomb field using the matrix method. It seems just possible, though unlikely, that the formula is not applicable to an electron in a field differing widely from a Coulomb field and that in such cases the scalar product ls which occurs in the formula has a value different from $\frac{1}{2}[j(j-1)+l(l+1)-s(s+1)]$, which it has in a Coulomb field. If this were the case, the central field result would be expected to apply still to the core electrons (especially to the L doublet) for which the proportional deviation from a Coulomb field is not large, but not to the series electron.

Apart from this point, the general agreement of X-ray and optical term values calculated by the methods here given with those observed is very satisfactory.

§ 8. Solution on Wave Mechanics for the Case when the Equation for Apsidal Distances has four Roots

As already mentioned, Rb was chosen as the first atom for which to do numerical work, partly in order to examine what happened on the wave mechanics in the case in which the equation

$$2v - \varepsilon - l(l+1)/r^2 = 0$$

giving the apses for the orbital atomic model, has four roots, which for Rb occurs when $l = 2$.

In Fig. 4 are shown three curves of $2v - l(l+1)/r^2$ as a function of r for different possible fields, a horizontal line being drawn at height ε, which is supposed to be a characteristic value for all three fields.

The usual curve of $2v - l(l+1)/r^2$ has one maximum, which may lie inside or outside the core; a horizontal line can then only cut it at two points, giving the apses of the corresponding orbit which is penetrating or non-penetrating respectively in the two cases. But the curve may have two maxima, in which case a horizontal line can cut it in four places, giving the apses of an internal and external orbit of the same energy. The curve for this case is drawn full, that for a penetrating orbit is drawn broken, and that for a non-penetrating orbit dot-and-dash. The ranges covered by the orbits in the various cases are marked below.

In Fig. 5 the curves of P against r are drawn to correspond with the curves in Fig. 4, the arbitrary constant being chosen so that the outer part of the curve in Fig. 5 is the same in all three cases. (The curves are diagrammatic only.)

From the differential equation for P

$$P'' + [2v - \varepsilon - l(l+1)/r^2]P = 0,$$

it follows that when $2v - \varepsilon - l(l+1)/r^2$ (i.e. the ordinate of the curve of $2v - l(l+1)/r^2$ above the line at height ε in Fig. 4) is positive, which is the case when r lies between the apses of the corresponding orbit, then the (P, r) curve is concave to the r axis, and when $2v - \varepsilon - l(l+1)$ is negative the (P, r) curve is convex to the r axis.

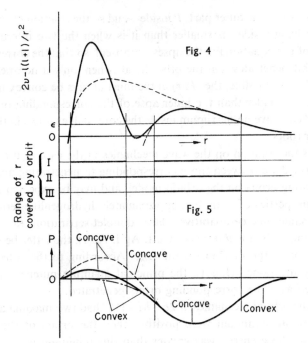

FIG. 4. Curves of $2v - l(l+1)/r^2$ as function of r.

Case I   (Internal and external orbit of same energy) ————

Case II   (Penetrating orbit) – – – – – –

Case III (Non-penetrating orbit) — · — · —

FIG. 5. Wave function corresponding to curves in FIG. 4. above. 'Convex' and 'concave' refer to curvature with respect to r axis. Figs. 4 and 5 to illustrate behaviour of wave function for field which gives separate internal and external orbits of same energy on orbital atomic model.

In the case of four roots, the internal and external orbits with the same energy are quite separate mechanically possible orbits (they would in general not both be quantum orbits), whereas on the wave mechanics both are included in a single solution of the wave equation, the complete separation being replaced by a range for r for which the (P, r) curve is convex to the r axis. The effect of this convex piece is that, for the same outer part, P inside—and so the fraction of the total charge lying inside—is smaller than it is when the two intermediate roots of the equation for the apses are absent, so that the corresponding orbit penetrates. On the other hand, when only a non-penetrating orbit is possible, the (P, r) curve must always be convex to the r axis for r smaller than the inner apse of the corresponding orbit, so that P^2 can have no maximum inside the core such as it has in the case of four roots.

Thus the solution on the wave mechanics in the case of four roots is intermediate between those corresponding to penetrating and non-penetrating orbits on the orbital model, and may be expected to give some properties of both, e.g. approximately hydrogenlike terms and at the same time comparatively large doublet separations (as shown, for example, by the d terms of Cu I, Ag I, Au I), the latter being due to the maximum of P inside the core. According to the convention already mentioned (I, § 1), the principal quantum number must be assigned as if the corresponding orbits penetrated.

On the orbital mechanics, if $2v - l(l+1)/r^2$ had two maxima and the intermediate minimum were positive, then the orbits of the series electron whose energy was greater than this minimum would correspond to non-penetrating orbits, and those whose energy was less would correspond to penetrating orbits, and quite a sharp break in the progression of the quantum defect along the sequence of terms would be expected in such a case*. On the wave mechanics, where the sharp separation between internal and external orbits of the same energy in the case of four roots has disappeared, no such sharp break in a sequence of terms would be expected. As the energy decreased, the range of r for which (P, r) curve is convex to the P axis would shrink and disappear, but that is all; an abnormal increase in quantum defect throughout the sequence, and possibly in doublet separation for the first few terms, would be expected, just as observed, for example, in the d terms of Al I, but no sharp break.

---

* See G. Wentzel, *Zeit. f. Phys.*, Vol. XIX, p. 52 (1923).

TABLE V SODIUM$^+$. APPROXIMATE SELF-CONSISTENT FIELD AND DISTRIBUTION OF CHARGE

r atomic units	Z	$-dZ/dr$ electrons per atomic unit	r atomic units	Z	$-dZ/dr$ electrons per atomic unit
0·00	11·00	0·0	0·5	7·15	8·8
0·02	10·98	2·8	0·6	6·26	9·0
0·04	10·90	7·0	0·7	5·37	8·5
0·06	10·72	10·2	0·8	4·55	7·6
0·08	10·49	11·9	0·9	3·84	6·5
0·10	10·25	12·1	1·0	3·25	5·39
0·12	10·02	11·4	1·2	2·39	3·47
0·14	9·80	10·5	1·4	1·82	2·16
0·16	9·61	9·3	1·6	1·47	1·31
0·18	9·43	8·2	1·8	1·27	0·77
0·20	9·27	7·3	2·0	1·16	0·45
0·25	8·94	6·0	2·5	1·04	0·12
0·30	8·64	6·0	3·0	1·01	0·03
0·35	8·33	6·6			
0·40	7·98	7·4			

§ 9. Sodium and Chlorine

Approximations to the self-consistent field have been worked out for the ions Na$^+$ and Cl$^-$ in a similar way to that for Rb$^+$, the immediate object being to determine the charge distribution, from which to calculate the functions F entering in the formula for the intensity of reflection of X-rays from crystals, in connection with recent experimental work on rocksalt. A comparison of the results of this work with the results of calculation, using the charge distribution found by the method discussed in this paper, will be given elsewhere, but it may be mentioned here that the result of the comparison is very satisfactory indeed.

In Tables V and VI the values of Z and the radial charge density $-dZ/dr$ for Na$^+$ and Cl$^-$ are given for the final field of the last approxi-

mation made, for which the maximum difference between initial and
final field was 0·05 for Na^+ and 0·08 for Cl^-. The values for Cl^- for
$r > 1$ are somewhat uncertain, as the main contribution is from the
outermost group of electrons, which is very sensitive to the changes in

TABLE VI CHLORINE⁻. APPROXIMATE SELF-CONSISTENT FIELD AND
DISTRIBUTION OF CHARGE

r atomic units	Z	$-dZ/dr$ electrons per atomic unit	r atomic units	Z	$-dZ/dr$ electrons per atomic unit
0·00	17·00	0·0	0·5	9·16	12·5
0·01	16·99	2·8	0·6	8·15	8·0
0·02	16·95	8·0	0·7	7·50	5·1
0·03	16·84	13·2	0·8	7·08	3·6
0·04	16·69	16·7	0·9	6·76	3·2
			1·0	6·43	3·27
0·06	16·32	19·6			
0·08	15·93	18·7	1·2	5·72	3·90
0·10	15·58	16·2	1·4	4·91	4·21
0·12	15·28	14·0	1·6	4·08	4·06
0·14	15·01	12·9	1·8	3·30	3·72
0·16	14·76	12·7	2·0	2·61	3·25
0·18	14·47	13·3			
0·20	14·19	14·3	2·5	1·30	2·05
			3·0	0·50	1·26
			3·5	0·00	0·80
			4·0	−0·33	0·52
0·25	13·42	17·0			
0·30	12·51	18·8	5	−0·69	0·23
0·35	11·57	18·7	6	−0·85	0·11
0·40	10·67	17·2	7	−0·93	0·05
			8	−0·97	0·02

the initial field. This sensitiveness is a result of the negative charge on
the ion, and made the work much more troublesome than for neutral
atoms or positive ions. For a multiply charged negative ion, for which
there is a repulsive field on one electron when far enough removed

from the rest, the distribution of charge for the outer electrons would be still more sensitive to the initial field, and the calculations would probably become unmanageable.

§ 10. Summary

The methods of solution of the wave equation for a central field given in the previous paper are applied to various atoms. For the core electrons, the details of the interaction of the electrons in a single n_k group are neglected, but an approximate correction is made for the fact that the distributed charge of an electron does not contribute to the field acting on itself (§ 2).

For a given atom the object of the work is to find a field such that the solutions of the wave equation for the core electrons in this field (corrected as just mentioned for each core electron) give a distribution of charge which reproduces the field. This is called the self-consistent field, and the process of finding it is one of successive approximation (§ 3).

Approximations to the self-consistent field have been found for He (§ 4), Rb⁺ (§ 5), Na⁺, Cl⁻ (§ 9). For He the energy parameter for the solution of the wave equation for one electron in the self-consistent field of the nucleus and the other corresponds to an ionisation potential of 24·85 volts (observed 24·6 volts); this agreement suggests that for other atoms the values of the energy parameter in the self-consistent field (corrected for each core electron) will probably give good approximations to the X-ray terms (§ 4).

The most extensive work has been carried out for Rb⁺. The distribution of charge given by the wave functions in the self-consistent field is compared with the distribution calculated by other methods (§ 6). The values of X-ray and optical terms calculated from the self-consistent field show satisfactory agreement with those observed (§ 7).

The wave mechanical analogue of the case in which on the orbital model an internal and an external orbit of the same energy are possible is discussed (§ 8).

2

The Calculation of Atomic Fields

L. H. Thomas (*Proc. Camb. Phil. Soc.* **23**, p. 542–548)

Trinity College

[*Received* 6 November, *read* 22 November 1926.]

The theoretical calculation of observable atomic constants is often only possible if the effective electric field inside the atom is known. Some fields have been calculated to fit observed data* but for many elements no such fields are available. In the following paper a method is given by which approximate fields can easily be determined for heavy atoms from theoretical considerations alone.

1. Assumptions and the deduction from them of an equation

The following assumptions are made.

(1) Relativity corrections can be neglected.

(2) In the atom there is an effective field given by potential V, depending only on the distance r from the nucleus, such that

$$V \to 0 \quad \text{as} \quad r \to \infty,$$

$$Vr \to E, \text{ the nuclear charge, as } r \to 0.$$

(3) Electrons are distributed uniformly in the six-dimensional phase space for the motion of an electron at the rate of two for each h^3 of

---

* D. R. Hartree, *Proc. Camb. Phil. Soc.*, 21, p. 625; E. Fues, *Zeit. für. Phys.*, 11, p. 369.

(six) volume. (This means one for each unit cell in the phase space of translation and rotation of a spinning electron.) The part of the phase space containing electrons is limited to that for which the orbits are closed.

(4) The potential V is itself determined by the nuclear charge and this distribution of electrons.

In reality the effective field at any point depends on whether the point is empty or occupied by a foreign electron or one or another atomic electron and on the circumstances of that occupation. These fields can only be expected to be sensibly the same or approximately calculable from the above assumptions if the density of electrons is large, that is, in the interior of heavy atoms.

If e, m, p are the charge, mass and momentum of an electron, the Hamiltonian function for the electronic motion is ((1) and (2) above),

$$\frac{1}{2m} p^2 - eV.$$

There are ((3) above) electrons at two for each h^3 of phase space for which

$$p < (2meV)^{\frac{1}{2}}$$

i.e. at

$$\frac{2}{h^3} \frac{4}{3} \pi (2meV)^{\frac{3}{2}}$$

per unit of ordinary (coordinate) space.

Thus ((4) above)

$$\nabla^2 V = 4\pi e \cdot \frac{2}{h^3} \frac{4}{3} \pi (2meV)^{\frac{3}{2}}$$

i.e.

$$\frac{1}{r^2} \frac{d}{dr} r^2 \frac{dV}{dr} = 4\pi e \frac{2}{h^3} \frac{4}{3} \pi (2me)^{\frac{3}{2}} V^{\frac{3}{2}} \qquad (1.1)$$

with ((2) above)

$$V \to 0 \quad \text{as} \quad r \to \infty,$$
$$Vr \to E \quad \text{as} \quad r \to 0.$$

Now express distance in terms of the 'radius of the normal orbit of the hydrogen atom,' $a = h^2/4\pi^2 m e^2 = 5\cdot3 \times 10^{-9}$ cms., potential in terms of the potential of an electron at this distance, so

$$r = \varrho \frac{h^2}{4\pi^2 m e^2},$$

$$V = \psi e \left| \frac{h^2}{4\pi^2 m e^2}, \right.$$

and equation (1.1) becomes

$$\frac{1}{\varrho^2} \frac{d}{d\varrho} \left(\varrho^2 \frac{d\psi}{d\varrho} \right) = \frac{8\sqrt{2}}{3\pi} \psi^{\frac{3}{2}} \tag{1.2}$$

with $\qquad \psi \to 0 \quad$ as $\quad \varrho \to \infty,$

$\varrho\psi \to N$, the atomic number, as $\varrho \to 0.$

(It is useful to note that with 'a' as unit of length, the charge and mass of the electron as units of charge and mass, $h = 2\pi$, whence (1.2) is at once verified.)

The 'effective nuclear charge' 'at distance ϱ is then given by

$$Z = -\varrho^2 \frac{d\psi}{d\varrho}.$$

Putting $\psi = (9\pi^2/128)\phi$, the equation for ϕ is

$$\frac{1}{\varrho^2} \frac{d}{d\varrho} \left(\varrho^2 \frac{d\phi}{d\varrho} \right) = \phi^{\frac{3}{2}} \tag{1.3}$$

2. Discussion of the equation

Write $\log \varrho = x$, $\varrho^4 \phi = w$, and the equation becomes

$$\frac{d^2 w}{dx^2} - 7 \frac{dw}{dx} + 12w = w^{\frac{3}{2}} \tag{2.1}$$

or, if $dw/dx = p$

$$\frac{dp}{dw} = 7 + \frac{w\left(w^{\frac{1}{2}} - 12\right)}{p} \tag{2.2}$$

The maximum and minimum locus of this equation is

$$p = -\frac{w\left(w^{\frac{1}{2}} - 12\right)}{7}.$$

The inflexion locus is

$$p = -\frac{2w\left(w^{\frac{1}{2}} - 12\right)}{7 \mp \left(1 + 6w^{\frac{1}{2}}\right)^{\frac{1}{2}}} = f(w),$$

and

$$\left(\frac{dp}{dw}\right)_{p = f(w)} - f'(w) = \pm 3w^{\frac{1}{2}}\left(w^{\frac{1}{2}} - 12\right)\left\{7 \mp \left(1 + 6w^{\frac{1}{2}}\right)^{\frac{1}{2}}\right\}^{-2}\left(1 + 6w^{\frac{1}{2}}\right)^{-\frac{1}{2}},$$

gives the direction in which the solutions cross the inflexion locus.

There are two singular points, $w = 0$, $p = 0$; $w = 144$, $p = 0$.

At $w, p \to 0$, $(4w - p)^4 \sim c(p - 3w)^3$, $(w > 0)$,

at $w \to 144$, $p \to 0$, $(7 \cdot 772(w - 144) - p)^{7 \cdot 772}(\cdot 772(w - 144) + p)^{\cdot 772} \sim c$,

give the form of the solutions, c being arbitrary.

The dp/dw discriminant gives $p = 0$, and $w = 144$ or $\phi = 144\varrho^{-4}$ as a singular solution.

There is an approximate particular solution,

$$\left. \begin{aligned} p &= -\frac{4\lambda}{\sqrt{(12)}} w\left(w^{\frac{1}{4}} - 12^{\frac{1}{2}}\right) \\ w &= \frac{144}{(1 + e^{-\lambda x})^4} \end{aligned} \right\} \tag{2.21}$$

which satisfies

$$\frac{3}{5\lambda}\frac{dp}{dw} = \frac{12}{5} + \frac{3}{\lambda} + \frac{w\left(w^{\frac{1}{2}} - 12\right)}{p}.$$

The solutions of (2.2) lie roughly as in the sketch (Fig. 1), the arrows give the direction of increase of ϱ. The only solutions for which $\phi \to 0$

as $\varrho \to \infty$ and $\phi = O\ (1/\varrho)$ as $\varrho \to 0$ correspond to the solution through O and A in the sketch—(2.21) is an approximation to this solution*. Different values of the nuclear charge correspond to the

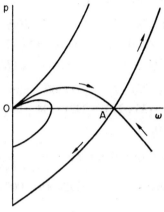

Fig. 1

replacement of x by $x+c$, which does not affect (2.1) so that if the equation is integrated numerically, starting from an initial position with w and p near A and any value of x, all the required solutions can be deduced.

3. The numerical integration

For the initial values put

$$w = 144(u+1)^{-4}, \quad v = \frac{du}{dx},$$

in (2.1), obtaining

$$(u+1)v\frac{dv}{du} - 5v^2 - 7(u+1)v - 3u(u+2) = 0.$$

---

* It is only at A that ϱ becomes infinite.

If $\quad v' = \lambda(u + au^2),$

$$G \equiv (u+1)v'\frac{dv'}{du} - 5v'^2 - 7(u+1)v' - 3u(u+2)$$

$$= u(\lambda^2 - 7\lambda - 6) + u^2(-4\lambda^2 - 7\lambda - 3 + a(3\lambda^2 - 7\lambda))$$
$$+ u^3(a\{-7\lambda^2 - 7\lambda\} + 2a^2\lambda^2) - 3a^2\lambda^2 u^4$$

$$= \tfrac{1}{2}u^2\{35\sqrt{(73)} - 292 + a(134 - 14\sqrt{(73)})\}$$
$$+ au^3\{7(4\sqrt{(73)} - 34) + a(61 - 7\sqrt{(73)})\} - 3a^2\lambda^2 u^4 .$$

For $\quad \lambda = -\tfrac{1}{2}(\sqrt{(73)} - 7) = -\cdot77200 \quad$ and $\quad u > 0.$

$G < 0 \quad$ for $\quad a = -(35\sqrt{(73)} - 292)/(134 - 14\sqrt{(73)}) = -\cdot0027900,$

$G > 0 \quad$ for $\quad a = 0,$

from which it can be shown that for $u > 0$

$$-\cdot77200u < \frac{du}{dx} < -\cdot77200(u - \cdot0027900u^2) \qquad (3.1)$$

For the actual numerical integration it is convenient to put

$$x = X\log_e 10, \quad w = 144 \times 10^Y,$$

so that (2.1) becomes

$$\frac{d^2Y}{dX^2} = \log_e 10\left\{12 \times 10^{\frac{1}{2}Y} + \cdot25 - \left(3\cdot5 - \frac{dY}{dX}\right)^2\right\} \qquad (3.2)$$

while

$$\psi = \frac{9\pi^2}{128}144 \times 10^{Y - 4(X + c)} \qquad (3.31)$$

$$Z = \frac{9\pi^2}{128}\left(4 - \frac{dY}{dX}\right)144 \times 10^{Y - 3(X + c)} \qquad (3.32)$$

$$\varrho = 10^{X + c} \qquad (3.33)$$

where c is to be determined from the atomic number. Z is the effective nuclear charge.

If $\qquad Y = -1, \quad u = 10^{\frac{1}{4}} - 1 = \cdot77828,$

and $\qquad\qquad 1\cdot3515 > \dfrac{dY}{dX} > 1\cdot3489 \qquad\qquad$ (from 3.1)

Starting with

$$X = 0, \quad Y = -1, \quad \frac{dY}{dX} = 1{\cdot}35,$$

numerical integration was carried out by steps of $\cdot 1$ to $X = -3$ by the aid of the formulae

$$\left(\frac{dy}{dx}\right)_{n+1} - \left(\frac{dy}{dx}\right)_{n} = \left(\frac{d^2y}{dx^2}\right)_{n} + \frac{1}{2} \varDelta \left(\frac{d^2y}{dx^2}\right)_{n-1} + \frac{5}{12} \varDelta^2 \left(\frac{d^2y}{dx^2}\right)_{n-2} + \cdots,$$

$$y_{n+1} - y_n = \left(\frac{dy}{dx}\right)_{n} + \frac{1}{2}\left(\frac{d^2y}{dx^2}\right)_{n} + \frac{1}{6} \varDelta \left(\frac{d^2y}{dx^2}\right)_{n-1}$$

$$+ \frac{1}{8} \varDelta^2 \left(\frac{d^2y}{dx^2}\right)_{n-2} + \cdots *.$$

For $X = -3$ it appears that

$$3{\cdot}5 - \frac{dY}{dX} = {\cdot}508, \quad \log_{10} 144 + Y = 7{\cdot}4385$$

so equation (3.32) gives

$$Z = \frac{9\pi^2}{128} \times 1{\cdot}008 \times 10^{2{\cdot}4385 - 3c},$$

i.e. $c = {\cdot}7611 - \frac{1}{3} \log_{10} N,$

since here closely enough $Z = N$ the atomic number.

e.g.   for $N = 55$ (caesium), $c = {\cdot}1810.$

4. Numerical results

The following table gives the values of

$$3{\cdot}5 - \frac{dY}{dX} \quad \text{and} \quad \log_{10} 144 + Y$$

---

* See Whittaker and Robinson, *The Calculus of Observations*, p. 365.

$-X$	$3\cdot5 - \dfrac{dY}{dX}$	$\log_{10} 144 + Y$	ϱ_0	Z_0	ψ_0	Z_1
0	2·150	1·1584	1·517	7·6	1·887	9·9
·1	2·015	1·0167	1·205	10·4	3·412	12·5
·2	1·880	·8614	·9572	13·7	6·008	16·0
·3	1·746	·6927	·7603	17·5	10·23	19·7
·4	1·615	·5105	·6040	21·6	16·90	24·3
·5	1·489	·3156	·4800	25·8	27·10	29·0
·6	1·371	·1086	·3811	30·1	42·26	33·4
·7	1·261	$\bar{1}$·8901	·3027	34·2	64·18	36·6
·8	1·160	$\bar{1}$·6611	·2404	38·0	95·15	39·5
·9	1·069	$\bar{1}$·4225	·1910	41·3	138·0	42·2
1·0	·987	$\bar{1}$·1752	1·517	44·2	196·1	44·7
1·1	·914	$\bar{2}$·9202	·1205	47·7	273·9	46·6
1·2	·851	$\bar{2}$·6584	·09572	48·7	376·4	47·8
1·3	·795	$\bar{2}$·3906	·07603	50·3	510·4	48·4
1·4	·747	$\bar{2}$·1176	·06040	51·5	683·8	49·3
1·5	·706	$\bar{3}$·8402	·04800	52·5	906·8	50·6
1·6	·671	$\bar{3}$·5590	·03811	53·2	1198	51·6
1·7	·642	$\bar{3}$·2746	·03027	53·8	1556	52·4
1·8	·614	$\bar{4}$·9875	·02404	54·0	2018	53·4
1·9	·595	$\bar{4}$·6979	·01910	54·4	2601	53·9
2·0	·577	$\bar{4}$·4064	·01517	54·6	3340	54·1
2·1	·564	$\bar{4}$·1134	·01205	54·8	4273	54·4
2·2	·552	$\bar{5}$·8191	·009572	54·9	5450	54·6
2·3	·542	$\bar{5}$·5238	·007603	54·9	6936	54·7
2·4	·534	$\bar{5}$·2276	·006040	55·0	8809	54·8
2·5	·527	$\bar{6}$·9306	·004800	55·0	11170	
2·6	·521	$\bar{6}$·6330	·003811	55·0	14140	
2·7	·517	$\bar{6}$·3349	·003027	55·0	17870	
2·8	·513	$\bar{6}$·0364	·002404	55·0	22580	
2·9	·510	$\bar{7}$·7376	·001910	55·0	28570	
3·0	·508	$\bar{7}$·4385	·001517	55·0	35960	

found by numerical integration and the corresponding values of ϱ, Z, ψ for caesium. The former may be in error by about 10 in the last decimal place.

For $\varrho_0 < \cdot006$, the field is sensibly a Coulomb field.

For $\varrho_0 > 1\cdot5$, the approximate formula (2.21) is an accurate enough solution of the differential equation, but this equation is not an accurate representation of the facts.

For the element of atomic number N the corresponding values are given by

$$\varrho = \varrho_0 \left(\frac{55}{N} \right)^{\frac{1}{3}},$$

$$Z = Z_0 \left(\frac{N}{55} \right),$$

$$\psi = \psi_0 \left(\frac{N}{55} \right)^{\frac{4}{3}}.$$

The values Z_1 are (unpublished) values calculated by Mr Hartree for caesium from the observed levels and which he has very kindly allowed me to include for comparison.

In conclusion, I wish to thank Professor Bohr and Professor Kramers for their encouragement when I was carrying out the numerical integration last March.

3

A Statistical Method for the Determination of Some Atomic Properties and the Application of this Method to the Theory of the Periodic System of Elements

E. FERMI (*Zeits. für Physik*, **48**, p. 73–79)

Rome

[With 1 illustration. *Received* 23 February 1928.]

In a heavy atom the electrons can be regarded as a kind of atmosphere surrounding the nucleus, which exists in a state of complete degeneracy. One can calculate approximately the distribution of electrons around the nucleus by means of a statistical analysis; this is applied to the theory of the formation of shells of electrons in the atom. The agreement with experiment is satisfactory.

§ 1. The object of the present work is to set out some findings concerning the distribution of electrons in a heavy atom, which may be arrived at by means of a statistical consideration of the electrons in the atom; in other words, in this work the electrons are regarded as an electron gas surrounding the nucleus.

One can easily establish that the density of this gas is so great that at temperatures attainable in practice it exists permanently in a state of complete degeneracy, so that for the purpose of the calculation proposed by the author, one must apply statistics based upon the Pauli Exclusion Principle. By this means both the distribution of electrons around the nucleus and their velocity distribution become temperature independent, provided that the temperatures are not too high, as is the case in practice.

In § 2 we shall calculate the density of the electron gas as a function of the distance r from the nucleus and of the atomic number Z; from this the increase of electrical potential within the atom will naturally also be determined. The number of electrons in the atom having a given angular momentum and thus a given azimuthal quantum number k will be calculated in § 3. It follows that electrons with a given azimuthal quantum number in the normal atomic state first appear at a known value of Z; these values of Z indicate the positions in the Periodic System for which an anomaly in the system begins. In § 4 the theoretical results are compared with experiment throughout the Periodic System.

§ 2. In order to calculate the electron distribution, we must first establish the relationship between the density n of the electrons and the electrical potential V. The potential energy of an electron is $-eV$; according to classical statistics the density would thus be proportional to $e^{eV/kT}$. According to statistics based on the Pauli Exclusion Principle, the resulting relationship between the density and the potential assumes the following form (one should at the same time take into account the double statistical weight factor of the electrons):

$$n = 2 \frac{(2\pi mkT)^{\frac{3}{2}}}{h^3} F(\alpha e^{eV/kT}), \tag{1}$$

where α represents a constant. The function F in our case (complete degeneracy) has the following asymptotic form:

$$F(A) = \frac{4}{3\sqrt{\pi}} (\log A)^{\frac{3}{2}}.$$

From this we obtain

$$n = \frac{2^{\frac{9}{2}} \pi m^{\frac{3}{2}} e^{\frac{3}{2}}}{3h^3} v^{\frac{3}{2}}, \tag{2}$$

where

$$v = V + \frac{kT}{e} \log \alpha, \tag{3}$$

and thus, except for an additional constant, v represents the potential.

Now the charge density is $-ne$, so that we have for the potential the following differential equation:

$$\Delta v = 4\pi n e = \frac{2^{\frac{13}{2}}\pi^2 m^{\frac{3}{2}} e^{\frac{5}{2}}}{3h^3} \, v^{\frac{3}{2}}. \tag{4}$$

Since v in this case is a function of r, the distance from the nucleus, the previous equation becomes

$$\frac{d^2v}{dr^2} + \frac{2}{r}\frac{dv}{dr} = \frac{2^{\frac{13}{2}}\pi^2 m^{\frac{3}{2}} e^{\frac{5}{2}}}{3h^3} \, v^{\frac{3}{2}}. \tag{5}$$

Now the potential in the vicinity of the nucleus, where the electron screening becomes negligible, is equal to Ze/r, so that one has:

$$\lim_{r=0} rv = Ze. \tag{6}$$

Furthermore one has

$$\int n \, d\tau = Z,$$

$d\tau$ = the volume element, since the total number of electrons is Z. If one substitutes expression (2) for n in this equation, and $d\tau = 4\pi r^2 \, dr$, one has

$$\frac{2^{\frac{13}{2}}\pi^2 m^{\frac{3}{2}} e^{\frac{5}{2}}}{3h^3} \int_0^\infty v^{\frac{3}{2}} r^2 \, dr = Ze. \tag{7}$$

The potential v is thus determined by the differential equation (5) together with the conditions (6) and (7). One can simplify these equations by substituting for r and v two other variables,

$$x = r/\mu \quad \text{and} \quad \psi = v/\gamma, \tag{8}$$

where

$$\mu = \frac{3^{\frac{2}{3}}h^2}{2^{\frac{13}{3}}\pi^{\frac{4}{3}}me^2 Z^{\frac{1}{3}}}, \quad \gamma = \frac{2^{\frac{13}{3}}\pi^{\frac{4}{3}}me^3 Z^{\frac{4}{3}}}{3^{\frac{2}{3}}h^2} \tag{9}$$

are constants. Equations (5), (6) and (7) then become

$$\left.\begin{array}{c} \dfrac{d^2\psi}{dx^2} + \dfrac{2}{x}\dfrac{d\psi}{dx} = \psi^{\frac{3}{2}}, \\[2mm] \lim_{x=0} x\psi = 1, \\[2mm] \displaystyle\int_0^\infty \psi^{\frac{3}{2}} x^2\, dx = 1. \end{array}\right\} \tag{10}$$

These equations can be simplified by putting

$$\phi = x\psi; \tag{11}$$

one then obtains

$$\frac{d^2\phi}{dx^2} = \frac{\phi^{\frac{3}{2}}}{\sqrt{x}}, \tag{12}$$

$$\phi(0) = 1, \qquad \int_0^\infty \phi^{\frac{3}{2}} \sqrt(x)\, dx = 1. \tag{13}$$

It is easy to see that the last condition is automatically satisfied when ϕ vanishes for $x = \infty$. One therefore requires a solution of (12) subject to the boundary conditions $\phi(0) = 1$, $\phi(\infty) = 0$.

By means of a numerical process, I have obtained the desired solution in the following table.

x	$\varphi(x)$	x	$\varphi(x)$	x	$\varphi(x)$
0·0	1·000	1·5	0·315	10	0·024
0·1	0·882	2·0	0·244	11	0·020
0·2	0·793	2·5	0·194	12	0·017
0·3	0·721	3·0	0·157	13	0·014
0·4	0·660	3·5	0·130	14	0·012
0·5	0·607	4	0·108	15	0·011
0·6	0·562	5	0·079	16	0·009
0·7	0·521	6	0·059	17	0·008
0·8	0·485	7	0·046	18	0·007
0·9	0·453	8	0·037	19	0·006
1·0	0·425	9	0·029	20	0·005

In this way the electrical potential in the interior of the atom is determined completely. Explicitly one has

$$v = \gamma \frac{\phi(x)}{x} = \gamma \mu \frac{\phi(x)}{r} = \frac{Ze}{r} \phi\left(\frac{r}{\mu}\right). \tag{14}$$

From (2) the electron density distribution is then also determined; one obtains

$$n = \frac{2^{\frac{9}{2}}\pi m^{\frac{3}{2}} Z^{\frac{3}{2}} e^3}{3h^3} \frac{1}{r^{\frac{3}{2}}} \varphi^{\frac{3}{2}}\left(\frac{r}{\mu}\right). \tag{15}$$

§ 3. In order to be able to apply these results to the theory of the Periodic Table, we must at this point solve the following problem:

How many electrons in the atom possess an angular momentum between p and $p+dp$?

In order to provide an answer to this question, we need to know the velocity distribution of electrons at every point within the atom. Now let n be the density of electrons at one point such that our gas is completely degenerate, i.e. such that the cells in phase space (per unit volume) which are associated with the n lowest energy states, are completely occupied and the remaining cells are completely vacant.

Now it is well known that the number of cells per unit volume of momentum space (momentum space = phase space per unit volume of normal space) equals $1/h^3$, since, however, electrons have a statistical weighting factor of 2, the number of electronic states per unit volume of momentum space is $2/h^3$. The number of states which correspond to a momentum of less than P_0, is thus

$$\frac{2}{h^3} \frac{4\pi}{3} P_0^3.$$

From this it follows that the distribution of the n points representing n electrons in momentum space is as follows:

The points are uniformly distributed with constant density $2/h^3$ inside the sphere whose centre corresponds to zero momentum and whose radius P_0 can be calculated from

$$\frac{2}{h^3} \frac{4\pi}{3} P_0^3 = n \qquad (16)$$

Outside this sphere there are no points.

The number of electrons per unit volume whose momentum component normal to the radius vector r lies between P and $P+dP$ is thus

$$\frac{2}{h^3} 2\pi P \, dP . 2 \sqrt{(P_0^2 - P^2)} = \frac{8\pi}{h^3} P \sqrt{(P_0^2 - P^2)} \, dP$$

$$= \frac{8\pi}{h^3} P \sqrt{\left[\left(\frac{3h^2 n}{8\pi} \right)^{\frac{2}{3}} - P^2 \right]}.$$

The angular momentum of these electrons, however, lies between the limits $p = rP$ and $p+dp = (P+dP)r$. We thus find that the number of electrons per unit volume whose angular momentum lies between p and $p+dp$ is

$$\frac{8\pi}{h^3} \frac{p}{r} \sqrt{\left[\left(\frac{3h^2 n}{8\pi} \right)^{\frac{2}{3}} - \frac{p^2}{r^2} \right]} \frac{dp}{r}.$$

If we multiply this expression by $4\pi r^2 \, dr$ and integrate over all values of r for which the radicand has real values, we obtain the number dN_p of electrons in the entire atom whose angular momentum lies between p and $p+dp$. We thus have

$$dN_p = \frac{32\pi^2}{h^3} p \, dp \int \sqrt{\left[\left(\frac{3h^3 n}{8\pi} \right)^{\frac{2}{3}} - \frac{p^2}{r^2} \right]} \, dr.$$

If we substitute expression (15) for n, we obtain

$$dN_p = \frac{32\pi^2}{h^3} p \, dp \int \sqrt{\left[2mZe^2 r\phi\left(\frac{r}{\mu} \right) - p^2 \right]} \frac{dr}{r}.$$

Instead of p we now introduce the azimuthal quantum number k where $p = kh/2\pi$; we thus find that the number of electrons whose azimuthal quantum number lies between k and $k+dk$ is as follows:

$$dN_k = \frac{8}{h} k \, dk \int \sqrt{\left[2mZe^2r\phi\left(\frac{r}{\mu}\right) - \frac{h^2k^2}{4\pi^2} \right]} \frac{dr}{r}.$$

Again, in place of r, we introduce the variable $x = r/\mu$, when we obtain

$$dN_k = \left(\frac{48}{\pi^2}\right)^{\frac{1}{3}} Z^{\frac{1}{3}} k \, dk \int \sqrt{\left[x\phi(x) - \frac{2^{\frac{4}{3}}k^2}{3^{\frac{2}{3}}\pi^{\frac{2}{3}}Z^{\frac{2}{3}}} \right]} \frac{dx}{x}.$$

Again, the integral should include all values of x for which the radicand has real values.

We now introduce the following function:

$$\Phi(A) = \int \sqrt{[x\Phi(x) - A]} \frac{dx}{x}.$$

We then have

$$dN_k = \left(\frac{48}{\pi^2}\right)^{\frac{1}{3}} Z^{\frac{1}{3}} k \, dk \Phi\left(\frac{2^{\frac{4}{3}}k^2}{3^{\frac{2}{3}}\pi^{\frac{2}{3}}Z^{\frac{2}{3}}} \right).$$

I have calculated the function $\Phi(A)$ numerically: it vanishes for $A > 0\cdot49$ (because in this case the radicand has no real value.) For $A < 0\cdot49$ one can obtain its value from the following table:

A	$\Phi(A)$	A	$\Phi(A)$
0·49	0·00	0·2	1·48
0·4	0·36	0·1	2·2
0·3	0·88	0·0	3·2

Now we know from quantum theory that k can only assume discrete values and, although according to the new quantum theory, $k = 0, 1, 2, \ldots$, it is nevertheless well known that one obtains the best agreement with experiment if one assumes half integral values of k; i.e. $k = \frac{1}{2}, \frac{3}{2}, \frac{5}{2}, \frac{7}{2}, \ldots$ for the s-, p-, d-, f-, $\ldots$ electrons. Now since our entire calculation was carried out using the old quantum theory, we must insert half integral values for k and for dk, of course, 1. We thus have finally that the number of electrons with a specified value of k in an atom with atomic number Z is given by the following formula:

$$N_k = \left(\frac{48}{\pi^2}\right)^{\frac{1}{3}} Z^{\frac{1}{3}} k \Phi\left(\frac{2^{\frac{4}{3}} k^2}{3^{\frac{2}{3}} \pi^{\frac{2}{3}} Z^{\frac{2}{3}}}\right). \tag{17}$$

§ 4. We want now to turn to the comparison of our theoretical results with experiment. It is well known that the atoms in the Periodic Table are built up in accordance with the rule that the Z-electrons occupy the Z most stable quantum states of the atom. If the electron screening were very small, all states with smaller principal quantum number n would be more stable than those with larger n, and the filling-up of the shells in the Bohr–Stoner table would proceed without any gaps. As is well known, however, the ordering is influenced by the screening effect to the extent that the sequence of energies of the electronic levels is no longer that of the principal quantum number, and it can happen that electrons with smaller n and larger k are more weakly bound than are electrons with larger n and smaller k. This provides an explanation for the well known gaps in the Bohr–Stoner table; e.g. the $4f$ states are first filled in the rare earth elements, for which the $6s$ states are already occupied.

Our theory allows these points to be studied in an approximately quantitative manner. The agreement with experiment is illustrated in the figure. In this, four pairs of curves are drawn; each pair consists of a continuous and a zig-zag line and corresponds to a value of the azimuthal quantum number. (The four pairs refer to the s-, p-, d- and f-electrons, i.e. to $k = \frac{1}{2}, \frac{3}{2}, \frac{5}{2},$ and $\frac{7}{2}$.) The continuous curve shows the number N_k of the k-electrons as a function of Z according to the theory (eqn. (17)), the zig-zag line records the empirical values of N_k as given by the Stoner table. Of course the statistical nature of our theory precludes the possibility of its reproducing the precise

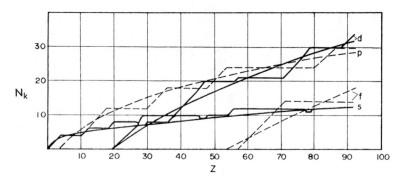

characteristics of the empirical curves. The general form of the curves is, nevertheless, depicted by the theory with considerable accuracy.

We notice in particular that one can calculate from the theory the values of the atomic number at which the *s*-, *p*-, *d*- and *f*-electrons begin to appear in the atom. These numbers are those for which the ordinates of the four theoretical curves assume the value of unity, i.e. 1, 5, 21, and 55 respectively. Now the *s*-electrons, in fact, already appear in hydrogen ($Z = 1$); the *p*-electrons first appear in boron ($Z = 5$); the *d*-electrons in scandium ($Z = 21$), where the first principal periodic anomaly begins; and the *f*-electrons in the element cerium ($Z = 58$), where the anomalous group of rare earths begins. One thus sees that the theory allows atomic numbers for which the various anomalies of the Periodic Table first appear to be predicted with a fair degree of accuracy.

I hope in a further article on the application of the statistical method to be able to investigate other atomic properties.

4

A Simplification of the Hartree–Fock Method

J. C. SLATER (*Phys. Rev.* **81**, p. 385–390)

Massachusetts Institute of Technology, * *Cambridge, Massachusetts*

[*Received* 28 September, 1950]

It is shown that the Hartree–Fock equations can be regarded as ordinary Schrödinger equations for the motion of electrons, each electron moving in a slightly different potential field, which is computed by electrostatics from all the charges of the system, positive and negative, corrected by the removal of an exchange charge, equal in magnitude to one electron, surrounding the electron whose motion is being investigated. By forming a weighted mean of the exchange charges, weighted and averaged over the various electronic wave functions at a given point of space, we set up an average potential field in which we can consider all of the electrons to move, thus leading to a great simplification of the Hartree–Fock method, and bringing it into agreement with the usual band picture of solids, in which all electrons are assumed to move in the same field. We can further replace the average exchange charge by the corresponding value which we should have in a free-electron gas whose local density is equal to the density of actual charge at the position in question; this results in a very simple expression for the average potential field, which still behaves qualitatively like that of the Hartree–Fock method. This simplified field is being applied to problems in atomic structure, with satisfactory results, and is adapted as well to problems of molecules and solids.

* The work described in this paper was supported in part by the Signal Corps, the Air Materiel Command, and the ONR, through the Research Laboratory of Electronics of M.I.T.

I. Introduction

THE Hartree–Fock equations[1] furnish the best set of one-electron wave functions for use in a self-consistent approximation to the problem of the motion of electrons in the field of atomic nuclei. However, they are so complicated to use that they have not been employed except in relatively simple cases. It is the purpose of the present paper to examine their meaning sufficiently closely so that we can see physically how to set up a simplification, which still preserves their main features. This simplified method yields a single potential in which we can assume that the electrons move, and we shall show the properties of this field for problems not only of single atoms but of molecules and solids, showing that it leads to a simplified self-consistent method for handling atomic wave functions, easy enough to apply so that we can look forward to using it even for heavy atoms.

II. The Hartree–Fock equations and their meaning

It is well known that the Hartree equations are obtained by varying one-electron wave functions $u_1(x)$, $u_2(x)$, ... $u_n(x)$, in such a way as to make the energy $\int u_1^*(x_1) \ldots u_n^*(x_n) H u_1(x_1) \ldots u_n(x_n) dx_1 \ldots dx_n$ an extreme, where H is the energy operator of a problem involving n electrons in the field of certain nuclei, and where the functions u_i are required to be normalized. Similarly the Hartree-Fock equations, as modified by Dirac,[2] are obtained by varying the u_i's so as to make the energy

$$
\frac{1}{n!} \int
\begin{vmatrix} u_1^*(x_1) \ldots u_1^*(x_n) \\ \cdots\cdots\cdots\cdots \\ u_n^*(x_1) \ldots u_n^*(x_n) \end{vmatrix}
H
\begin{vmatrix} u_1(x_1) \ldots u_1(x_n) \\ \cdots\cdots\cdots\cdots \\ u_n(x_1) \ldots u_n(x_n) \end{vmatrix}
dx_1 \ldots dx_n
$$

[1] J. C. Slater, *Phys. Rev.* **35**, 210 (1930); V. Fock, *Z. Physik* **61**, 126 (1930); L. Brillouin, *Les Champs Self-Consistents de Hartree et de Fock, Actualités Scientifiques et Industrielles* No. 159 (Hermann et Cie., 1934); D. R. Hartree and W. Hartree, *Proc. Roy. Soc.* **A150**, 9 (1935); and many other references.
[2] P. A. M. Dirac, Proc. Cambridge Phil. Soc. **26**, 376 (1930).

an extreme, where in this latter expression the u's are assumed to be functions depending on coordinates and spin, and where the integrations over the dx's are interpreted to include summing over the spins. The Hartree-Fock equations can then be written in the form

$$H_1 u_i(x_1) + \left[\sum_{k=1}^{n} \int u_k^*(x_2)\, u_k(x_2)\, (e^2/4\pi\varepsilon_0 r_{12})\, dx_2 \right] u_i(x_1)$$

$$- \sum_{k=1}^{n} \left[\int u_k^*(x_2)\, u_i(x_2)\, (e^2/4\pi\varepsilon_0 r_{12})\, dx_2 \right] u_k(x_1) = E_i u_i(x_1). \quad (1)$$

Here H_1 is the kinetic energy operator for the electron of coordinate x_1, plus its potential energy in the field of all nuclei; $e^2/4\pi\varepsilon_0 r_{12}$ is the Coulomb potential energy of interaction between electrons 1 and 2, expressed in mks units; to get the corresponding formula in Gaussian units we omit the factor $4\pi\varepsilon_0$, and to get it in atomic units we replace $e^2/4\pi\varepsilon_0$ by 2. The u_i's as before are assumed to depend on spin as well as coordinates, and the integrations over dx_2 include summation over spin, so that the exchange terms, the last ones on the left side of eq. (1), automatically vanish unless the functions u_i and u_k correspond to spins in the same direction.

The Hartree-Fock equations in the form given present an appearance which seems to differ from the ordinary one-electron type of Schrödinger equation, and for this reason it is ordinarily thought that they cannot be given a simple physical interpretation. This assumption arises partly from the paper of Dirac[2], in which they are interpreted in a rather involved way. The second term on the left of (1) is simple: it is clearly the Coulomb potential energy, acting on the electron at position x_1, of all the electronic charge, including that of the ith wave function whose wave equation we are writing. The last term on the left, the exchange term, however, is peculiar, in that is is multiplied by $u_k(x_1)$ rather than by $u_i(x_1)$. It must somehow correct for the fact that the electron does not act on itself, which it would be doing if this term were omitted. In the Hartree, as opposed to the Hartree-Fock,

15*

equations, this is obvious. There the last term differs from that in the Hartree-Fock equations only in that all terms in the summation are omitted except the ith; the exchange term in that case then merely cancels the term in $k = i$ from the Coulomb interaction found in the second term. The main point of our discussion is to show that an equally simple interpretation of this term can be given in the Hartree-Fock equations.

Let us first state in words what the interpretation proves to be; then we can more easily describe the way in which the equations lead to it. We can subdivide the total density of all electrons into two parts, ϱ_+ from those with plus spins, ϱ_- from those with minus spins; the two together add to the quantity $-e\Sigma(k = 1 \ldots n)u_k^*(x)u_k(x)$, where e is the magnitude of the electronic charge. Then we can show that the Hartree-Fock eq. (1) for a wave function u_i which happens to correspond to an electron with a plus spin is an ordinary Schrödinger equation for an electron moving in a perfectly conventional potential field. This field is calculated by electrostatics from all the nuclei, and from a distribution of electronic charge consisting of the whole of ϱ_-, but of ϱ_+ corrected by removing from the immediate vicinity of the electron, whose wave function we are finding, a correction or exchange charge density whose total amount is just enough to equal a single electronic charge. That is, this corrected charge distribution equals the charge of $n-1$ electrons, as it should. The exchange charge density equals just ϱ_+ at the position of the electron in question, gradually falling off as we go away from that point. We can get a rough idea of the distance in which it has fallen to a small value by replacing it by a constant density ϱ_+ inside a sphere of radius r_0, zero outside the sphere. We have $\frac{4}{3}\pi r_0^3|\varrho_+| = e$, or

$$r_0 = (3e/4\pi|\varrho_+|)^{\frac{1}{3}}. \tag{2}$$

The situation is then much as if the corrected charge density equaled the actual total electronic charge density outside this sphere, but was only ϱ_- within the sphere; there is a sort of hole, sometimes called the

Fermi[3] or exchange hole, surrounding the electron in question, consisting of a deficiency of charge of the same spin as the electron in question. Actually, of course, this exchange hole does not have a sharp boundary, but the charge density of the same spin as the electron in question gradually builds up as we go away from this electron. Similar statements hold for the field acting on an electron of minus spin.

The exchange hole clearly is different for wave functions of the two spins, provided ϱ_+ and ϱ_- are different; examination proves further that it is different for each different wave function u_i. It is this difference which leads to the complicated form of the Hartree-Fock equation; and the simplification which we shall introduce in a later section is that of using sort of an averaged exchange hole for all the electrons. The difference between the exchange charge for two wave functions u_i corresponding to the same spin is not great, however. We have already seen that the radius r_0 which we obtain by assuming a hole of constant density depends only on ϱ_+ (for a plus spin), and hence is the same for all u_i's of that spin. Thus the exchange holes for different u_i's of the same spin will only show small differences. We shall later examine these differences for the case of a free electron gas, and show that they are really not large. It is this small dependence on u_i which will make it reasonable to use an averaged exchange charge in the simplified method which we shall suggest later.

To agree with the qualitative description which we have just given, we then expect the exchange charge density at point x_2, producing a field acting on the electron at x_1 whose wave function $u_i(x_1)$ we are determining by the Hartree-Fock eq. (1), to integrate over dx_2 to $-e$ (a single electronic charge), and to be equal when x_2 approaches x_1

---

[3] E. Wigner and F. Seitz, *Phys. Rev.* **43**, 804 (1933); ibid. **46**, 509 (1934). The discussion of Wigner and Seitz was one of the first to show a proper understanding of the main points taken up in the present paper, which must be understood to represent a generalization and extension of previously suggested ideas, rather than an entirely new approach. See also L. Brillouin, *J. de Phys et le Radium*, **5**, 413 (1934) for a discussion somewhat similar to the present one.

to the quantity

$$-e \sum_{\substack{k=1 \\ \text{spin } k = \text{spin } i}}^{n} u_k^*(x_1)\, u_k(x_1). \tag{3}$$

We shall now show that this is the case.

To show it, we rewrite (1) in the equivalent form[4]

$$H_1 u_i(x_1) + \left[\sum_{k=1}^{n} \int u_k^*(x_2)\, u_k(x_2)\, (e^2/4\pi\varepsilon_0 r_{12})\, dx_2 \right] u_i(x_1)$$

$$- \left[\sum_{k=1}^{n} \frac{\displaystyle\int u_i^*(x_1)\, u_k^*(x_2)\, u_k(x_1)\, u_i(x_2)\, (e^2/4\pi\varepsilon_0 r_{12})\, dx_2}{u_i^*(x_1)\, u_i(x_1)} \right] u_i(x_1) = E_i u_i(x_1). \tag{4}$$

The exchange term now appears as the product of a function of x_1, times the function $u_i(x_1)$; thus it has the standard form of a potential energy term in a one-electron Schrödinger equation. This exchange potential energy is the potential energy, at the position of the first electron, of the exchange charge density,

$$-e \sum_{k=1}^{n} \frac{u_i^*(x_1)\, u_k^*(x_2)\, u_k(x_1)\, u_i(x_2)}{u_i^*(x_1)\, u_i(x_1)}, \tag{5}$$

located at the position x_2 of the second electron. We note as we expect that the exchange charge density depends on the position of the first electron, as well as the second, and also on the quantum state i in which this first electron is located. We note, furthermore, that the total charge is that of a single electron. To show this, we integrate the exchange charge density (5) over dx_2, and find at once, on account of the orthogonality of the u_i's (which follows from the Hartree-Fock equations) that the integral over all space is $-e$. Furthermore, as x_2 approaches x_1, we see at once that the exchange charge density approaches the correct value (3), where the restriction that the spins of i

[4] J. C. Slater and H. M. Krutter, *Phys. Rev.* **47**, 559 (1935); particularly p. 564, where this same method is used in discussing the Thomas-Fermi method.

and k must be equal arises from (1), where an exchange term $u_k^*(x_2)u_i(x_2)$ is automatically zero unless this condition is satisfied. Thus we have shown that the exchange charge density (5) satisfies all the conditions necessary to justify our qualitative discussion of its behavior. In a later section, where we work out detailed values for the free-electron case, we can examine its properties more in detail.

The great difference between the Hartree and the Hartree-Fock methods is the fact that in the Hartree-Fock method the exchange hole or correction charge appropriate for an electron at x_1 moves around to follow that electron; in the Hartree method it does not, the correction charge depending only on the index i of the wave function u_i. If our problem is a single atom, this is not very important, but in a crystal, for instance a metal, the difference is profound. Thus consider a periodic lattice, in which the one-electron functions u_i are modulated plane waves, corresponding to $1/N$ of an electronic charge on each of the N atoms of the crystal. In the Hartree scheme, the potential acting on the electron in the wave function u_i is that of all electrons, minus this charge corresponding to $1/N$ of an electron on each atom. This correction charge is so spread out that its effect on the potential field is completely negligible, and each electron acts as if it were in the field of all electrons, thus finding itself in the field of a neutral atom when near any of the nuclei of the metal. On the other hand, with the Hartree-Fock equations, the exchange charge is located near the position x_1 of the electron in question, moving around with it, so that when this electron is on a given atom, the exchange charge is removed largely from that atom, leaving it in the form of a positive ion, which, as our physical intuition tells us, is the correct situation.

III. Averaged exchange charge

We have seen that the exchange charges for different wave functions u_i corresponding to the same spin are not very different from each other, since in every case they reduce to the same value when $x_2 = x_1$, and integrate to the same value over all space. Furthermore, in a system containing equal or approximately equal numbers of electrons with both spins, ϱ_+ and ϱ_- will be at least approximately the same, so that exchange charges for different u_i's even of opposite spins will be nearly the same. It then seems clear that we shall make no very great error if we use a weighted mean of the exchange charge density, weighting over i, for each value of x_1. The result of this will be that we shall have a single potential field to use for the Schrödinger equation for each u_i, simplifying greatly the application of the Hartree-Fock method. Let us first set up this average exchange charge and the consequent averaged exchange potential, then give some discussion of their properties and uses.

The probability that an electron at x_1 should be in the state i is evidently $u_i^*(x_1)\, u_i(x_1)/[\Sigma_j u_j^*(x_1)\, u_j(x_1)]$. We can then use this quantity as a weighting factor to weight the exchange charge density (5). When we do this, we find as the average exchange charge density the quantity[5]

$$-e\frac{\displaystyle\sum_{j=1}^{n}\sum_{k=1}^{n} u_j^*(x_1)\, u_k^*(x_2)\, u_k(x_1)\, u_j(x_2)}{\displaystyle\sum_{j=1}^{n} u_j^*(x_1)\, u_j(x_1)}. \tag{6}$$

Using this average exchange charge density, we come to the following Schrödinger equations for the u_i's, as substitutes for the Hartree-Fock equations:

$$H_1 u_i(x_1) + \left[\sum_{k=1}^{n} \int u_k^*(x_2)\, u_k(x_2)\, (e^2/4\pi\varepsilon_0 r_{12})\, dx_2 \right.$$

$$\left. -\frac{\displaystyle\sum_{j=1}^{n}\sum_{k=1}^{n} \int u_j^*(x_1)\, u_k^*(x_2)\, u_k(x_1)\, u_j(x_2)\, (e^2/4\pi\varepsilon_0 r_{12})\, dx_2}{\displaystyle\sum_{j=1}^{n} u_j^*(x_1)\, u_j(x_1)} \right] u_i(x_1) = E_i u_i(x_1). \tag{7}$$

[5] J. C. Slater, *Rev. Mod. Phys.* **6**, 209 (1934), particularly p. 267, where this same expression is used for similar purposes, but without pointing out that it is the weighted mean of the exchange charge density found in the Hartree-Fock equations.

The wave functions u_i, and energy values E_i, as determined from these equations, will not be quite so accurate as those determined from the Hartree-Fock equations; but they will at least be much better than those found from the Hartree equations, particularly for the case of the crystal, and they have the great advantage that they are all solutions of the same potential problem. This automatically brings one good feature, which the solutions possess in common with solutions of the Hartree-Fock equations, but which solutions of the Hartree equations do not have: the functions u_i are all orthogonal to each other.

There is one aspect of eqs. (7) which is very important. In the last few years there has been a great development of the energy-band theory of semiconductors. This is all based on the hypothesis that we can build up a model of a solid in which each electron moves independently in a potential field which is made up from the nuclei, and all other electrons except the one in question. The electric field derived from this potential is sometimes called the motive field acting on the electron. Each wave function corresponds to a definite energy level, and the Fermi statistics are applied to the distribution of the electrons in these levels. The soundest way to set up this potential acting on each electron is by the Hartree-Fock method, but we see by our present discussion that this implies a different potential energy or motive for each electron, or each u_i. If we wish to have a single motive field appropriate for all electrons, the best thing we can do is to use the weighted mean suggested in the present section. Thus eqs. (7) may well be taken to be the basis of the ordinary form of the energy-band theory of solids.

In many problems, we are interested in cases of degeneracy, not merely in evaluating the wave function of a single nondegenerate stationary state. Thus we may be solving a problem of multiplet structure in an atom or molecule, or discussing ferromagnetism in a solid. In such a case we start with a number of degenerate or approximately degenerate energy levels, corresponding to different orientations of orbit or spin, or in some cases (as in the hybridization of atomic orbitals) corresponding to different total or azimuthal quantum numbers, and then carry out perturbations. If we take the Hartree-Fock scheme literally, we shall use different potentials for finding the u_i's of each of these various unperturbed functions. It is highly desirable in such cases, in the interests of simplicity, to modify the method so as to use the same potential function for the calculation of each wave function. This may involve even more averaging than is contemplated in setting up eqs. (6) and (7). As one illustration, Hartree's use of a spherical potential for discussing atomic structure is an example of this procedure; this involves averaging over all orientations of the various orbital angular momenta of electrons which are not in closed shells. Whether we are using the Hartree scheme or the present simplification of the Hartree-Fock scheme, such averaging over orientations seems certainly desirable. Again, in studying ferromagnetism, the potentials to use, according to the scheme of the present paper, will depend on the net magnetization, or on the number of electrons of each spin. It is much simpler to handle such a problem, however, by using a single potential function, and that will usually be chosen to be that representing the unmagnetized state, with equal numbers of plus and minus spins.

In all these cases which we have just been discussing, we use one-electron wave functions which are slightly less accurate than those found by the Hartree-Fock scheme. When we apply perturbation methods, we must remember this, computing the matrix components of the exact energy operator with respect to these somewhat incorrect wave functions, remembering the wave equation (for instance (7)) which they actually satisfy. Nondiagonal matrix components of energy between these somewhat inaccurate wave functions will be somewhat larger than those between exact Hartree-Fock functions. Nevertheless they will still not be very large, for the wave functions are still quite accurate; the slight decrease in exactness is much more than made up by the simplicity of the method. The energy values computed by averaging the exact energy operator over the wave function will be very nearly the same as for Hartree-Fock functions, on account of the theorem that the mean value of energy over an incorrect wave function has errors only of the second order of small quantities.

IV. The exchange charge for the free-electron case

The calculations of exchange charge and exchange potential which we have been describing in general language can be carried out exactly for the case of a free-electron gas, as is well known. In this section we shall give the results, as an illustration of the general case. Then we shall point out in the next section that by using a free-electron approximation we can get an exchange potential much simpler than that of eq. (7), which still is accurate enough for many purposes.

Let us have a free-electron gas with n electrons in the volume V, half of them of each spin; the volume is assumed to be filled with a uniform distribution of positive charge, just enough to make it electri-

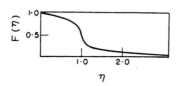

FIG. 1. $F(\eta)$ as function of η (from eq. (11)).

cally neutral. The electrons are assumed to obey the Fermi statistics. Then by elementary methods we find that they occupy energy levels with uniform density in momentum space, out to a level whose energy is $P^2/2m = (h^2/2m)(3n/8\pi V)^{\frac{2}{3}}$, corresponding to a maximum momentum $P = h(3n/8\pi V)^{\frac{1}{3}}$. The de Broglie wavelength

$$d = h/P = (8\pi V/3n)^{\frac{1}{3}} \tag{8}$$

associated with this maximum momentum is clearly related to the radius r_0 of the exchange hole, which we introduced in eq. (2). When we notice that $|\varrho_+|$ which appeared there equals $ne/2V$, we see that

$$d = (4\pi/3)^{\frac{2}{3}} r_0. \tag{9}$$

We can now state some of the principal results of the application of this model to the exchange charge density and exchange energy. The exchange potential energy can be conveniently stated in terms of the ratio $\eta = p/P$ of the magnitude of the momentum of the electron to the maximum momentum corresponding to the top of the Fermi distribution. It is[2]

$$\text{exchange potential energy} = (e^2/4\pi\varepsilon_0)\,(4P/h)\,F(\eta)$$

$$= (6/\pi)^{\frac{2}{3}}\,(e^2/4\pi\varepsilon_0 r_0)\,F(\eta), \tag{10}$$

where

$$F(\eta) = \frac{1}{2} + \frac{1-\eta^2}{4\eta}\,\ln\left[(1+\eta)/(1-\eta)\right]. \tag{11}$$

The function $F(\eta)$ is shown in Fig. 1. It goes from unity when $\eta = 0$, for an electron of zero energy, to $\frac{1}{2}$ when $\eta = 1$, at the top of the Fermi distribution. We see that this exchange potential energy is of the form which we should expect. If we had a sphere of radius r_0, filled with uniform charge density $|\varrho_+| = ne/2V$, the potential energy of an electronic charge at the center of the sphere would be $\frac{3}{2}(e^2/4\pi\varepsilon_0 r_0)$, while the value from eq. (10) is $1\cdot54(e^2/4\pi\varepsilon_0 r_0)$ at the bottom of the Fermi band, half this value at the top. Thus this simple model of an exchange hole of constant charge density gives a qualitatively correct value for the exchange potential, and rather accurate quantitative value and the extreme difference between top and bottom of the band corresponds only to a factor of 2 in the exchange potential.

If now we average over-all wave functions, we find that the properly weighted average of $F(\eta)$ is $\frac{3}{4}$. Thus the exchange potential energy of the averaged exchange charge[6] is $(\frac{3}{4})\,(6/\pi)^{\frac{2}{3}}\,(e^2/4\pi\varepsilon_0 r_0)$. This can also be found from the averaged exchange charge density. This charge density is[3]

$$\frac{\varrho}{2}\left[\frac{3\sin(r/d)-(r/d)\cos(r/d)}{(r/d)^3}\right]^2, \tag{12}$$

[6] F. Bloch, *Z. Physik*, **57**, 545 (1929) gave the first derivation of this value.

where ϱ is the total charge density of electrons, d is given by eqs. (8) and (9), and r is the distance from point x_1, where the electron whose wave function we are computing is located, to x_2, where we are finding the exchange charge density. This function (12) is shown in Fig. 2, plotted as a function of r/r_0, and we see that it does in fact represent a

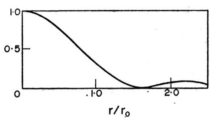

FIG. 2. Exchange charge density (divided by $\varrho/2$) plotted as a function of r/r_0, from eq. (12), where r_0 is given by eq. (9).

density which equals $\varrho/2$ when $r = 0$, and falls to small values at approximately $r = r_0$. The potential energy of an electron at the center of this averaged exchange charge distribution is just the value $\left(\frac{3}{4}\right)(6/\pi)^{\frac{2}{3}}(e^2/4\pi\varepsilon_0 r_0)$ previously given.

V. Use of the free-electron approximation for the exchange potential

From the argument of section III, it is clear that the exchange charge density (6), and the corresponding potential appearing in (7), must depend on the density of electronic charge, but not greatly on anything else. Thus in no case will we expect it to be very different from what we should have in a free-electron gas of the same charge density. We may then make a further approximation and simplification, beyond that of section III; we may approximate the averaged exchange potential by what we should have in a free-leectron gas of the same density, as given in section IV.[7] Thus, combining (10) and (2), we have

$$\text{exchange potential energy} = -\tfrac{3}{4}(6/\pi)^{\frac{2}{3}} (e^2/4\pi\varepsilon_0 r_0)$$

$$= -3(e^2/4\pi\varepsilon_0)(3n/8\pi V)^{\frac{1}{3}}, \qquad (13)$$

where we are now to interpret n/V as the local density of electrons, a function of position. If we recall that this is $\Sigma(k) u_k^*(x) u_k(x)$, we finally have as our simplified Schrödinger equation for the one-electron functions u_i, to replace (7),

$$H_1 u_i(x_1) + \left[\Sigma(k) \int u_k^*(x_2) u_k(x_2) (e^2/4\pi\varepsilon_0 r_{12}) dx_2 \right.$$

$$\left. -3(e^2/4\pi\varepsilon_0) \left\{ \frac{3}{8\pi} \Sigma(k) u_k^*(x_1) u_k(x_1) \right\}^{\frac{1}{3}} \right] u_i(x_1) = E_i u_i(x_1). \qquad (14)$$

This equation is in practice a very simple one to apply. We solve it for each of the wave functions u_i, then find the total charge density arising from all these wave functions, and can at once calculate the potential energy, including the exchange term, to go into (14), so as to check the self-consistency of the solution. Here, as before, we change to Gaussian units by omitting $4\pi\varepsilon_0$, and atomic units by changing $e^2/4\pi\varepsilon_0$ to 2.

One result of this formulation of the self-consistent problem is of immediate interest. In a periodic potential problem such as a crystal, it is obvious that the total charge density will have the same periodicity as the potential. Thus the corrected potential of eq. (14) will also be periodic, and hence the functions u_i will be modulated according to Bloch's theorem. In other words, such modulated functions are the only type which can follow from a proper application of our simplification of the Hartree-Fock method to a periodic potential problem.

---

[7] This method of treating the exchange potential as if the electrons formed part of a free-electron gas is similar to what is done in the Thomas–Fermi method with exchange (see Dirac (reference 2), Slater and Krutter (reference 4), and L. Brillouin, *L'Atome de Thomas–Fermi et la Méthode du Champ "Self-Consistent"*, *Actualités Scientifiques et Industrielles*, No. 160 (Hermann et Cie., 1934)).

Our general method is applicable to any problem of atoms, molecules, or solids. It is easy to give it a very explicit formulation for the case of atoms, which can then be used for the self-consistent treatment of atomic structure. Let the electrostatic potential of the nucleus, and of all electrons, at distance r from the nucleus of a spherical atom, be $Z_p(r)e/4\pi\varepsilon_0 r$. Then the charge density is given by Poisson's equation as $\varrho = -\varepsilon_0 \nabla^2(Z_p e/4\pi\varepsilon_0 r)$. When we express the Laplacian in spherical coordinates, this gives at once $\varrho = -(e/4\pi)(1/r)d^2Z_p/dr^2$. This is the quantity which is expressed as $-e\Sigma(k)u_k^*(x)u_k(x)$. Thus the exchange potential energy becomes $-3(e^2/4\pi\varepsilon_0)\left[\left(\frac{3}{32}\pi^2\right)(1/r)d^2Z_p/dr^2\right]^{\frac{1}{3}}$, and, finally, the total potential energy, for use in the Schrödinger equation for $u_i(x_1)$, is

$$-\frac{e^2}{4\pi\varepsilon_0 r}\left[Z_p+3\left(\frac{3}{32\pi^2}\right)^{\frac{1}{3}}\left(r^2\frac{d^2Z_p}{dr^2}\right)^{\frac{1}{3}}\right]. \tag{15}$$

To carry out a self-consistent solution for an atom, using this simplified method, we then find a Z_p such that the wave functions u_i, determined from a single Schrödinger equation using the potential energy (15), determined from Z_p, add to give a charge density which would lead by Poisson's equation to a potential energy $-e^2Z_p/4\pi\varepsilon_0 r$.

In order to check the applicability of the method Mr. George W. Pratt is carrying out a self-consistent solution of the ion Cu$^+$ by this method. This ion was chosen, as being the heaviest one for which solutions by both the Hartree and the Hartree-Fock methods are available for comparison. The solution has gone far enough to show that the wave functions and energy parameters E_i determined from it are not far from those found by the Hartree and the Hartree-Fock methods. The discrepancies come principally from large values of r, where the charge density is small, and our free-electron approximation for exchange is not very good. Over most of the range of r, however, the approximation seems very satisfactory. Detailed results will be reported later. The great advantages of this method for numerical calculation are clear from this example which has been worked out. Actual calculation is simpler than for the original Hartree scheme, since only one potential function need be computed, and can be used for all wave functions. The wave functions have the advantage of being orthogonal; and they possess a considerable part of the accuracy which the solutions of the Hartree-Fock equations possess, in contrast to the Hartree equations. It is to be hoped that they will make enough simplification so that it will be possible to carry out calculations for more complicated atoms than have yet been attempted by the Hartree-Fock method. At the same time the method should prove valuable in setting up solutions for molecules and solids.

Index